中国海洋经济统计年鉴

CHINA MARINE ECONOMIC STATISTICAL YEARBOOK

2018

自然资源部　编

Edited by

Ministry of Natural Resources of the People's Republic of China

海洋出版社

2020年·北京

图书在版编目(CIP)数据

中国海洋经济统计年鉴.2018/自然资源部编.—
北京:海洋出版社, 2019.11
ISBN 978-7-5210-0469-4

Ⅰ.①中… Ⅱ.①自… Ⅲ.①海洋经济-统计资料-
中国-2018-年鉴 Ⅳ.①P74-54

中国版本图书馆 CIP 数据核字(2020)第 003157 号

ZHONGGUO HAIYANG JINGJI TONGJI NIANJIAN 2018

责任编辑:王 溪
责任印制:赵麟苏
海洋出版社 出版发行
http://www.oceanpress.com.cn
北京市海淀区大慧寺路 8 号 邮编:100081
北京朝阳印刷厂有限责任公司印刷 新华书店北京发行所经销
2019 年 11 月第 1 版 2020 年 7 月第 1 次印刷
开本:889mm×1194mm 1/16 印张:17.25
字数:430 千字 定价:168.00 元
发行部:62132549 邮购部:68038093 总编室:62114335

《中国海洋经济统计年鉴》编委会

《中国海洋经济统计年鉴》编辑部

编 者 说 明

一、《中国海洋经济统计年鉴2018》系统收录了全国和沿海区域2017年开发、利用和保护海洋的各类活动以及社会经济概况的统计数据，是一部全面反映中国海洋经济发展有关情况的资料性年鉴，全书为中英文对照。

二、本年鉴所涉及的沿海区域包括沿海地区、沿海城市和沿海地带，按《沿海行政区域分类与代码》(HY/T 094—2006)的顺序排列。

三、本年鉴内容包括综合资料、海洋经济核算、主要海洋产业活动、主要海洋产业生产能力、海洋科学技术、海洋教育、海洋环境保护、海洋行政管理及公益服务、全国及沿海社会经济、部分世界海洋经济统计资料等十部分。

四、本年鉴根据《海洋统计报表制度》（国统制〔2017〕120号）和《海洋生产总值核算制度》（国统制〔2016〕165号），资料主要来源于沿海省（自治区、直辖市）统计局、自然资源（海洋）主管部门以及20个有关涉海部、局、总公司。

五、本年鉴中除特殊说明外，所有价值量指标均为当年价，年鉴中每部分后附有主要统计指标解释，对指标的含义、统计范围和统计方法做了简要说明。统计数据中的其他说明置于表的下方。有续表的资料，如有注释均置于第一张表的下方。

六、本年鉴中涉及的历史数据，均以本年鉴出版的最新数据为准；本年鉴中部分数据合计数或相对数由于单位取舍不同而产生的计算误差，均未做机械调整。

七、本年鉴表格中符号使用说明：“空格”表示该项统计指标数据不详或无该项数据；“#”表示其中的主要项；其他符号如“*”或“①”等表示本表后面有注释。

八、本年鉴资料国内部分未包括香港特别行政区、澳门特别行政区和台湾省数据。

九、《中国海洋经济统计年鉴》在编撰过程中，得到了各有关单位的大力支持，在此表示衷心的感谢。本年鉴中如有疏漏和不妥之处，敬请读者批评指正。

《中国海洋经济统计年鉴》编辑部

Editor's Notes

I. The *China Marine Economic Statistical Yearbook* (2018), which has included the statistical data on the national and coastal area's various activities of developing, utilizing and protecting the ocean as well as economy and society in 2017, is a data almanac reflecting in an all-round way China's marine economic development, and it is a Chinese-English bilingual edition.

II. The coastal areas covered by the Yearbook are the coastal regions, coastal cities and coastal zones, which are arranged in order according to the *Coastal Administrative Areas Classification and Codes* (HY/T 094-2006).

III. The data in the Yearbook consist of 10 sections, namely, integrated data, marine economic accounting, major marine industrial activities, production capacity of major marine industries, marine science and technology, marine education, marine environmental protection, marine administration and public-good service, national and coastal socioeconomy, and part of the world's marine economic statistics data.

IV. The Yearbook is based on the *Marine Statistics Report System* (Guotongzhi 〔2017〕 No.120) and the *Ocean Gross Product Accounting System* (Guotongzhi 〔2016〕 No. 165), its data mainly come from the statistical bureaus and departments in charge of natural resources (marine affairs) of the coastal provinces, autonomous regions, and municipalities directly under the Central Government as well as the 20 ocean-related ministries, bureaus and general corporations concerned.

V. Unless otherwise specified in the Yearbook, all the value indicators are given at the current price. Each section is attached by explanatory notes to the major marine statistical indicators, giving a brief explanation of the meaning, statistical range and statistical methods of the major marine statistical indicators. Other notes to the statistical data are listed below the tables. For the data with continued tables, annotations, if any, are put below the first table.

VI. All the historical data covered in the Yearbook are subject to the latest data published in the Yearbook; For part of the data in the Yearbook, the calculation errors on totals or relative figures due to the difference in the unit trade-offs have not been adjusted.

VII. The usage of symbols in the tables: "Blank" indicates that the data of the statistical index are unknown for the time being or that there are no such data available;

"#" indicates the major items of the table; Other symbols, such as "*" or "①", indicate "see footnotes below".

VIII. The domestic part of the Yearbook does not include the data from Hong Kong Special Administrative Region, Macau Special Administrative Region and Taiwan Province.

IX. In the course of editing *China Marine Economic Statistical Yearbook*, we enjoyed energetic support from the various departments concerned and we hereby extend our heartfelt thanks to them. Criticisms and comments are welcome from readers on the oversights and inappropriateness, if any, in the Yearbook.

Editorial Department of
China Marine Economic Statistical Yearbook

目　次
CONTENTS

2017年我国海洋经济发展综述

2017年，在以习近平同志为核心的党中央坚强领导下，各级海洋行政管理部门认真贯彻落实党的十九大提出的各项决策和中央经济工作会议精神，紧紧围绕党中央、国务院加快建设海洋强国的战略部署，主动引领海洋经济发展新常态，深入推进供给侧结构性改革，海洋经济实现稳中向好发展，结构调整持续深化，新旧动能转换加速形成，海洋经济正在向高质量发展不断迈进。

一、全国海洋经济发展概况

2017年，全国海洋生产总值76749.0亿元，比上年增长6.9%（除特殊注明外，增长率均按可比价计算），海洋生产总值占国内生产总值的9.4%，占沿海地区生产总值的16.6%。

二、主要海洋产业发展情况

2017年，我国海洋产业继续保持稳步增长。其中，主要海洋产业实现增加值31122.5亿元，比上年增长7.4%，占海洋生产总值的40.6%，滨海旅游业和海洋交通运输业仍占主导地位。

海洋第一产业　2017年，海洋渔业总体保持平稳增长，全年实现增加值4700.7亿元，比上年增长1.8%。在海洋渔业“转方式，调结构”背景下，海洋水产品产量3321.7万吨，比上年增长0.6%。其中，海水养殖产量持续增加，达到2000.7万吨，比上年增长4.5%；海洋捕捞产量有所降低，为1112.4万吨，比上年减少6.3%；远洋渔业产量为208.6万吨，比上年增长5.0%。海水养殖面积进一步减少，为208.4万公顷，比上年减少0.7%。远洋渔船数量达到2491艘，比上年减少0.3%；总功率达到255.2万千瓦，比上年增长6.1%。

海洋第二产业 2017年，海洋油气业受国内外市场需求和生产结构调整影响，效益出现下滑，全年实现增加值1145.2亿元，比上年下降0.4%；海洋原油产量4886.3万吨，比上年下降5.3%，海洋天然气产量139.5亿立方米，比上年增长8.3%。海洋矿业、海洋盐业、海洋化工业受市场需求下降和去库存影响，呈现负增长，全年分别实现增加值为65.2亿元、42.3亿元、1021.0亿元，比上年分别下降7.5%、7.5%、3.0%。海洋生物医药业快速增长，产业集聚逐渐形成，全年实现增加值389.1亿元，比上年增长12.3%。海洋电力业继续保持良好发展势头，海上风电项目加快推进，全年实现增加值151.7亿元，比上年增长18.8%。海水利用业稳步增长，应用规模逐渐扩大，全年实现增加值15.8亿元，比上年增长15.3%。海洋船舶工业受国内外市场需求影响，手持订单下降，船企开工不足，全年实现增加值1091.5亿元，比上年下降5.7%。海洋工程建筑业受投资放缓影响，增长下行压力显现，增速回落，全年实现增加值1846.4亿元，比上年增长0.6%。

海洋第三产业 2017年，航运市场逐步复苏，海洋交通运输业增长态势进一步增强，全年海洋交通运输业实现增加值6081.0亿元，比上年增长5.5%。沿海港口货物吞吐量90.6亿吨，比上年增长7.1%；国际标准集装箱吞吐量2.0亿标准箱，比上年增长3.1%。滨海旅游业继续保持较快发展，我国海洋旅游新业态与新模式不断涌现，海洋保护区、海洋公园建设逐渐改善滨海环境，滨海旅游持续升温，全年滨海旅游业实现增加值14572.5亿元，比上年增长16.0%。主要沿海城市接待入境旅游者人数4758.6万人次，比上年增长7.2%，其中港澳台入境游客占45.6%，是主要的客源市场。

三、区域海洋经济发展情况

2017年，环渤海、长江三角洲和珠江三角洲地区海洋经济继续保持平稳增长态势，前两个地区占全国海洋生产总值的比重略有下降，分别为31.9%、29.3%，珠江三角洲地区占比略有上升，为23.1%。

环渤海地区海洋生产总值24507.3亿元，占地区生产总值比重16.5%。海洋产业增加值15309.7亿元，海洋相关产业增加值9197.7亿元。滨海旅游业、海洋交通运输业、海洋渔业、海洋工程建筑业四个产业增加值位居前列，其增加值之和占该地区主要海洋产业增加值的87.3%。

长江三角洲地区海洋生产总值22469.5亿元，占地区生产总值比重13.4%。海洋产业增加值14030.6亿元，海洋相关产业增加值8438.9亿元。滨海旅游业、海洋交通运输业、海洋渔业和海洋船舶工业四个产业增加值位居前列，其增加值之和占该地区主要海洋产业增加值的91.7%。

珠江三角洲地区海洋生产总值17725.0亿元，占地区生产总值比重达19.8%。海洋产业增加值11671.0亿元，海洋相关产业增加值6054.0亿元。滨海旅游业、海洋交通运输业、海洋化工业和海洋工程建筑业四个产业增加值位居前列，其增加值之和占该地区主要海洋产业增加值的84.4%。

四、科技教育

2017年，海洋科研教育继续保持稳步发展，统计的海洋科研机构共159个，从业人员29089人，海洋科研机构承担课题21257项，发表海洋科技论文15872篇，出版海洋科技著作388种，专利授权数3288件，其中发明专利2242件，拥有发明专利总数10352件。开设海洋专业的高等院校达595个，专任教师数497589人。高等教育和中等职业教育海洋专业毕业生数分别为93707人和15363人；招生人数分别为81609人和11995人；在校人数分别为278758人和32158人。

五、海洋环境保护与防灾减灾

2017 年，我国近岸局部海域污染依然严重，近岸以外海域海水质量良好。夏季和秋季，我国管辖海域劣于第四类海水水质标准的海域面积分别为 33720 平方千米和 47310 平方千米，实施监测的河口、海湾、滩涂湿地、珊瑚礁、红树林和海草床等海洋生态系统中，处于健康、亚健康和不健康状态的海洋生态系统分别占 20%、70%和 10%。我国各类海洋灾害共造成直接经济损失 64.0 亿元。其中，风暴潮灾害造成直接经济损失 55.8 亿元，占总直接经济损失的 87%。最大面积超过 100 平方千米（含）的赤潮过程共 12 次。

六、海洋行政管理

2017 年，行政管理工作扎实推进，各项海洋工作取得明显成效。全年新增宗海数量 1831 本，新增宗海面积 168111.6 公顷；全年提供海洋数值预报服务（国家级）共 7663 次，各项海洋观测数据获得量共 16325.4MB，开展海洋调查项目 423 个；全年共有 41 项国家标准和 157 项行业标准通过立项审查，出版国家标准 37 项，行业标准 17 项。

Summary of China's Marine Economic Development in 2017

In 2017, under the firm leadership of the CPC Central Committee with Comrade Xi Jinping at the core, the marine administrative departments at all levels conscientiously implemented the policy decisions made by 19th National Congress of the Communist Party of China and the spirit of the Central Economic Work Conference, closely focused on the strategic deployment of the Party Central Committee and the State Council to speed up the construction of a maritime power, and actively guided the new normal development of the marine economy, and further pushed forward the structural reform of the supply side. The marine economy has achieved steady and sound development, sustained deepening of structural adjustment, and accelerated the formation of new and old kinetic energy conversion, so that it is continuously progressing towards high-quality development.

I. Survey of the National Marine Economic Development

In 2017, the national Gross Ocean Product amounted to 7674.90 billion yuan, 6.9% up from the previous year (Unless otherwise specified, the growth rate is calculated at the comparable price), accounting for 9.4% of the national GDP and 16.6% of the Gross Coastal Product. The number of people employed by the marine-related sectors in the country reached

36.568 million, 0.343 million more than that in the previous year.

II. Development of Major Marine Industries

In 2017, China's marine industry continued to keep a steady growth. Among others, the major marine industries effected a value added of 3112.25 billion yuan, 7.4% up from the previous year, accounting for 40.6% of the Gross Ocean Product, in which coastal tourism and marine communications and transport still occupied the leading position.

Primary marine industry

In 2017, the marine fishery maintained a stable growth on the whole, and effected a full-year value added of 470.07 billion yuan, registering an increase of 1.8% over the previous year. Against the background of the marine fishery's "transferring mode and adjusting structure", the output of marine aquatic products amounted to 33.217 million tons, 0.6% up from the previous year, among which, the mariculture yeild continued to grow, reaching 20.007 million tons, 4.5% up from the previous year; the marine fishing yield somewhat dropped, amounting to 11.124 million tons, 6.3% down from the previous year; the output of the deep-sea fishing was 2.086 million tons, 5.0% up from the previous year. The area of mariculture was further reduced, amounting to 2.084 million hm^2, decreasing by 0.7% as compared with that in the previous year. The number of ocean-going fishing vessels reached 2491, 0.3% down from the previous year; and the total power amounted to 2.552 million kW, 6.1% up from the previous year.

Secondary marine industry

In 2017, due to the influence of the domestic and foreign market demand and the adjustment of production structure, the benefits of the offshore oil and gas industry declined, the annual added value was 114.52 billion yuan, 0.4% down from the previous year; the output of offshore crude oil was 48.863 million tons, 5.3% down from the previous year, and the output of offshore natural gas was 13.95 billion m^3, 8.3% up from the previous year. The marine mining industry, sea salt industry and marine chemical industry affected by the decrease of market demand and the destocking, showed a negative growth. The annual added value was 6.52 billion yuan, 4.23 billion yuan and 102.10 billion yuan respectively, 7.5%, 7.5% and 3.0% down from the previous year respectively. With the rapid growth of marine biomedicine industry and the gradual formation of industrial agglomeration, the annual added value reached 38.91 billion yuan, 12.3% up from the previous year. The marine electric power industry continued to maintain a good momentum of development, and the offshore wind power projects were accelerated, the annual added value reaching 15.17 billion yuan, 18.8% up from the previous year. The seawater utilization industry grew steadily, its application scale gradually expanded, and the annual added value reached 1.58billion yuan, 15.3% up from the previous year. Affected by the market demand at home and abroad, the marine shipbuilding industry has seen a decline in the orders held by domestic and foreign enterprises, the shipbuilding enterprises were operating below capacity, and the annual added value was 109.15 billion yuan, 5.7% down from the previous year. Affected by the slowdown in investment, the marine engineering construction industry has seen downward pressure on growth, the growth rate has dropped, and the annual added value was 184.64 billion yuan, 0.6% up from the previous year.

Tertiary marine industry

In 2017, the shipping market gradually recovered, the growth trend of marine communications and transport industry was further strengthened, and the annual added value was 608.10 billion yuan, 5.5% up from the previous year. The cargo handling capacity of coastal ports was 9.06 billion tons, 7.1% up from the previous year; and the handling capacity of international standardized containers was 200 million TEU, 3.1% up from the previous year. The coastal tourism industry continued to maintain rapid development. New forms and modes of marine tourism continue to emerge in China. The construction of marine reserves and marine parks gradually improved the coastal environment. Coastal tourism continued to heat up. The annual added value of coastal tourism industry reached 1457.25 billion yuan, 16.0% up from the previous year. Major coastal cities received 47.586 million inbound tourists, 7.2% up from the previous year, of which those from Hong Kong, Macao and Taiwan accounted for 45.6%, which is the main tourist source market.

III. Development of Regional Marine Economy

In 2017, the marine economy in the Round-the-Bohai, Changjiang River Delta and Zhujiang River Delta regions continued to maintain a posture of steady growth, the proportions of the first two regions in the national Gross Ocean Product decreased slightly, accounting for 31.9% and 29.3% respectively, while that of the Pearl River Delta region increased slightly, accounting for 23.1%.

The Gross Ocean Product of the Round-the-Bohai region was 2450.73 billion yuan, accounting for 16.5% of the Gross Regional Product. The value added of marine industries was 1530.97 billion yuan and that of the marine-related industries 919.77 billion yuan. The four industries of coastal tourism, marine communications and transport industry, marine fishery and marine engineering architecture industry were among the front-runners, the sum of their values added accounting for 87.3% of the value added of the major marine industries in the region.

The Gross Ocean Product of the Changjiang River Delta region was 2246.95 billion yuan, accounting for 13.4% of the Gross Regional Product. The value added of the marine industries was 1403.06 billion yuan and that of the marine-related industries 843.89 billion yuan. The four industries of coastal tourism, marine communications and transport, marine fishery and marine shipbuilding ranked in the forefront of all industries, the sum of their values added accounting for 91.7% of the value added of the major marine industries in the region.

The Gross Ocean Product of the Zhujiang River Delta region was 1772.50 billion yuan, accounting for 19.8% of the Gross Regional Product. The value added of the marine industries was 1167.10 billion yuan and that of the marine related industries 605.40 billion yuan. The four industries of coastal tourism, marine communications and transport, marine chemical industry and marine engineering architecture were among the front-runners, the sum of their values added accounting for 84.4% of the value added of the major marine industries in the region.

IV. Science, Technology and Education

In 2017, the marine scientific research and education continued to develop steadily. The statistics number of marine scientific research institutions totaled 159 with 29089 employees; the number of marine scientific and technological projects undertaken by the marine scientific research institutions was 21257; 15872 marine scientific and technological papers and 388 kinds of marine scientific and technological works were published; and the number of patents authorized was 3288, of which 2242 were the ones for discovery, and the number owned of invention patents was 10352; the number of the institutions of higher learning which offer marine specialties amounted to 595, with 497589 full-time teachers in this regard; the number of graduates from marine specialties of the higher learning and the secondary vocational education was 93707 and 15363 respectively; the number of students enrolled was 81609 and 11995 respectively; and that in school was 278758 and 32158 respectively.

V. Marine Environmental Protection

In 2017, pollution was still serious in the local nearshore sea area of China, but the seawater quality beyond the nearshore sea area was good. In summer and autumn, the sea areas under China's jurisdiction with the seawater quality inferior to Class 4 were 33720 km^2 and 47310 km^2 separately. Among the marine ecosystems monitored such as estuaries, bays, tidal flat wetland, coral reef, mangrove and sea grass bed etc., those in the healthy, subhealthy and unhealthy state accounted for 20%, 70% and 10% respectively. The direct economic loss caused by various marine disasters in

China reached 6.40 billion yuan, of which, that caused by storm surge disasters amounted to 5.58 billion yuan, accounting for 87% of the total direct economic loss. There were 12 times of red tide process each with the largest area exceeding (including) 100 km^2.

VI. Marine Administration

In 2017, marine administration was carried forward soundly and various marine efforts obtained obvious results. The number of the sea parcels newly approved was 1831, the area of sea parcels newly approved reached 168111.6 hm^2; marine numerical forecast service (National Level) was provided on 7663 occasions, the data quantity from various marine observations totaled 16325.4 MB, and 423 marine survey projects were carried out; and a total of 41 items of national standard and 157 items of professional standard had been authorized and examined, and 37 items of national standard and 17 items of professional standard published.

图 1 全国海洋生产总值及三次产业构成

China's Gross Ocean Product and Three Industries Composition

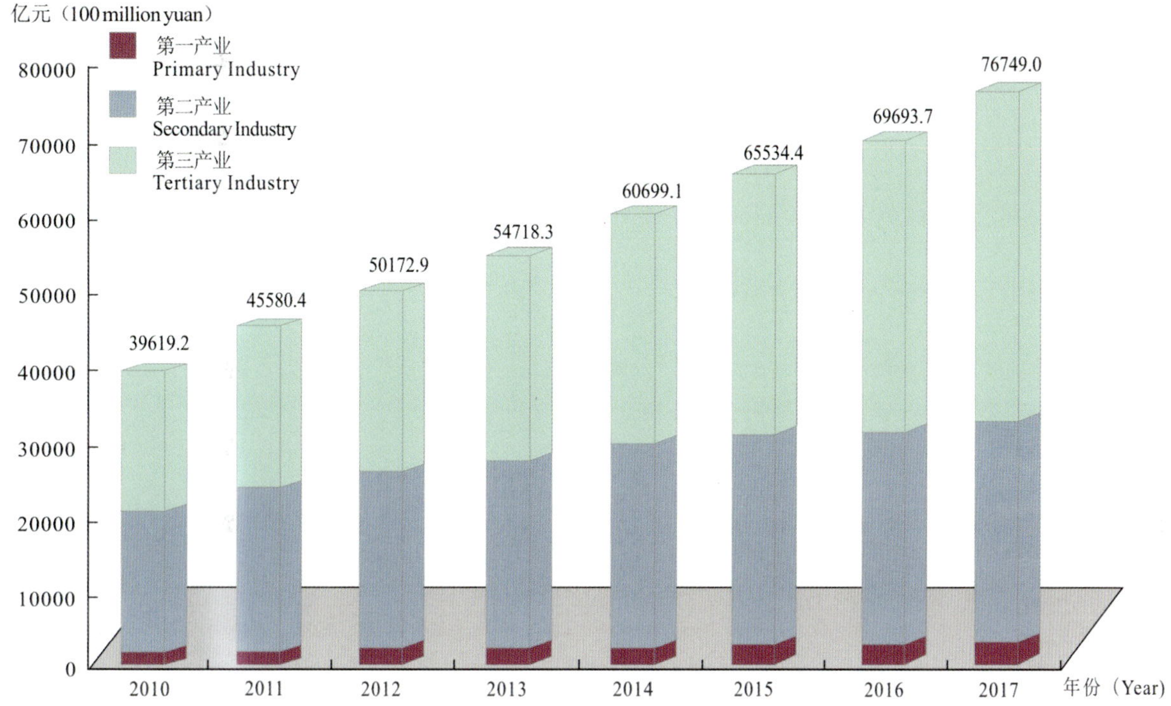

图 2 2017年全国主要海洋产业增加值构成

Composition of Added Values of the Major Marine Industries in China in 2017

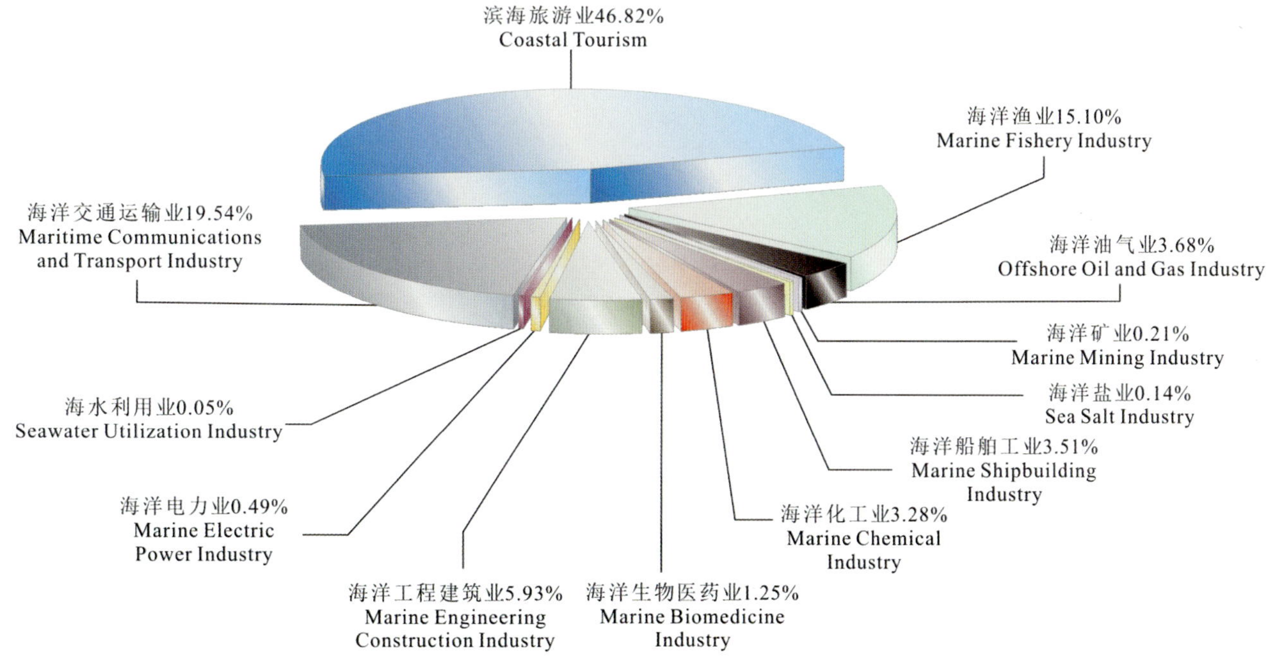

图 3 2017年沿海地区海洋生产总值

Gross Ocean Product by Coastal Regions in 2017

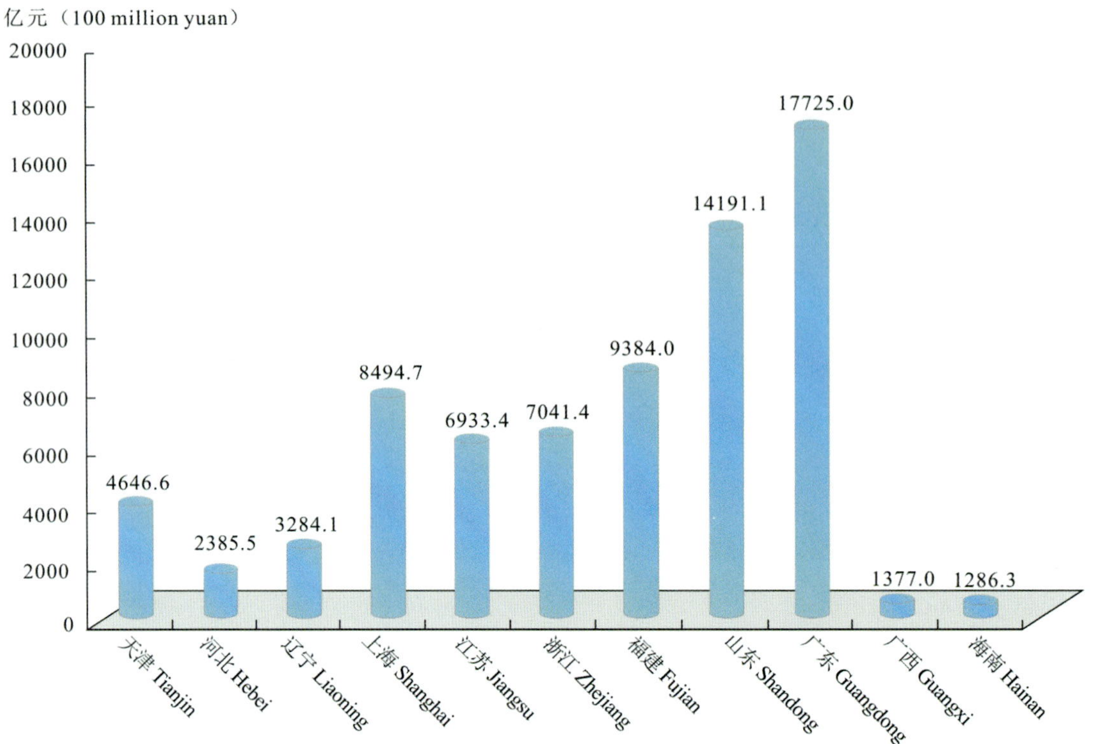

图 4 全国海洋捕捞和海水养殖产量

National Marine Catches and Mariculture Production

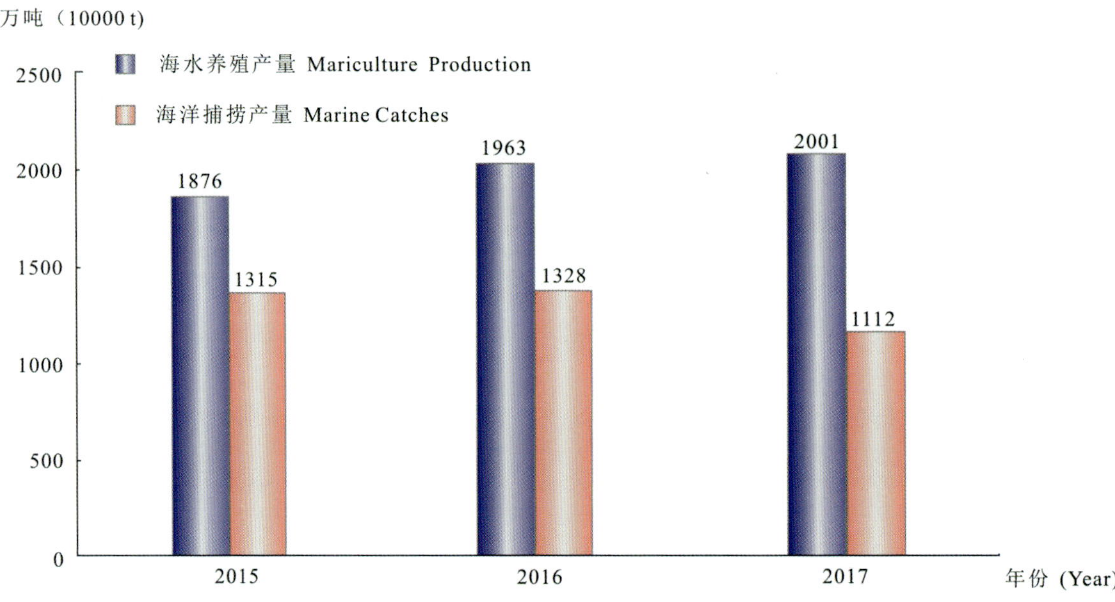

图 5 2017年沿海地区海洋捕捞和海水养殖产量
Marine Catches and Mariculture Production by Coastal Regions in 2017

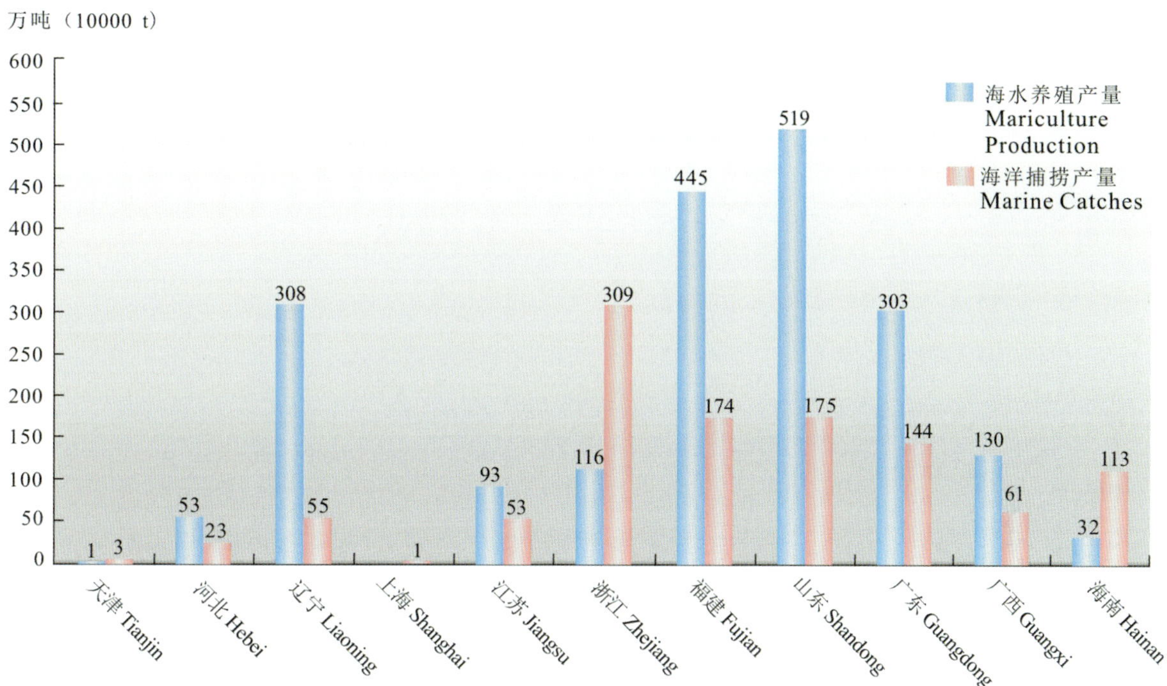

图 6 全国海洋石油和天然气产量
National Output of Offshore Oil and Natural Gas

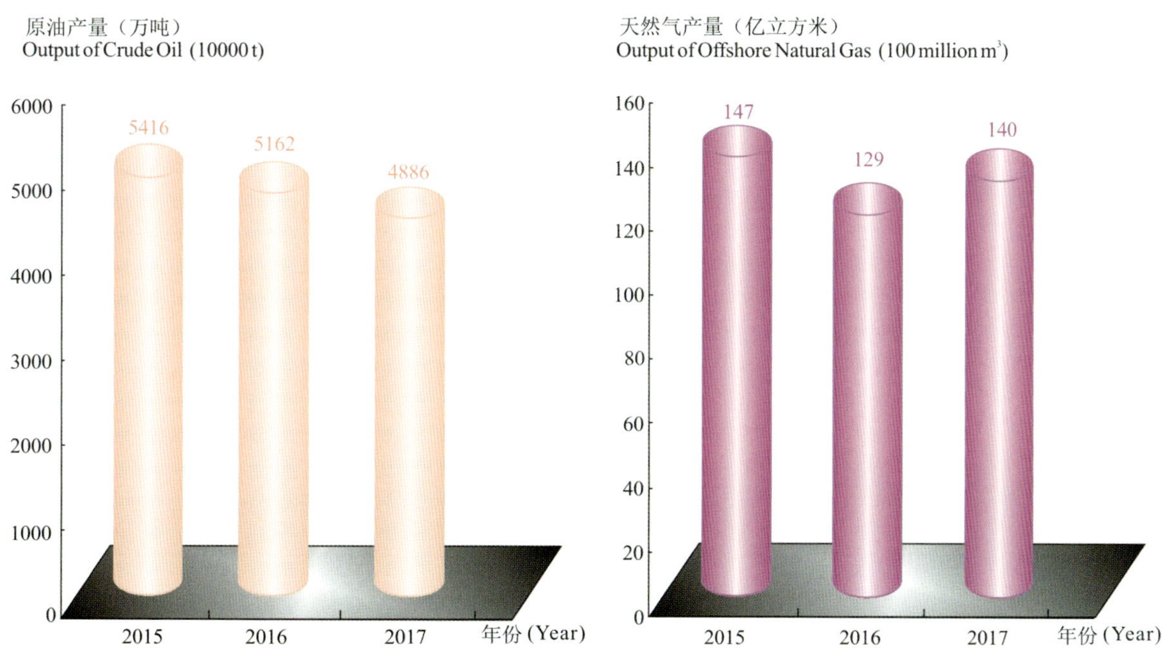

图 7 2017年沿海地区海洋原油产量

Offshore Crude Oil Production by Coastal Regions in 2017

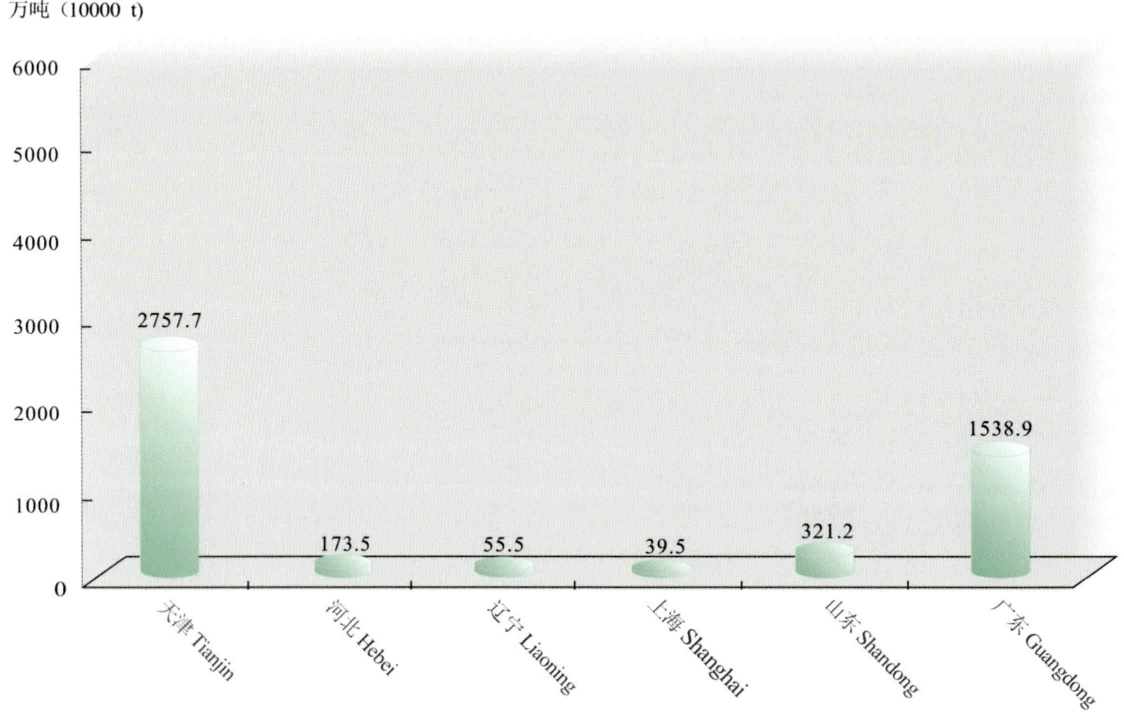

图 8 2017年沿海地区海洋天然气产量

Offshore Natural Gas Production by Coastal Regions in 2017

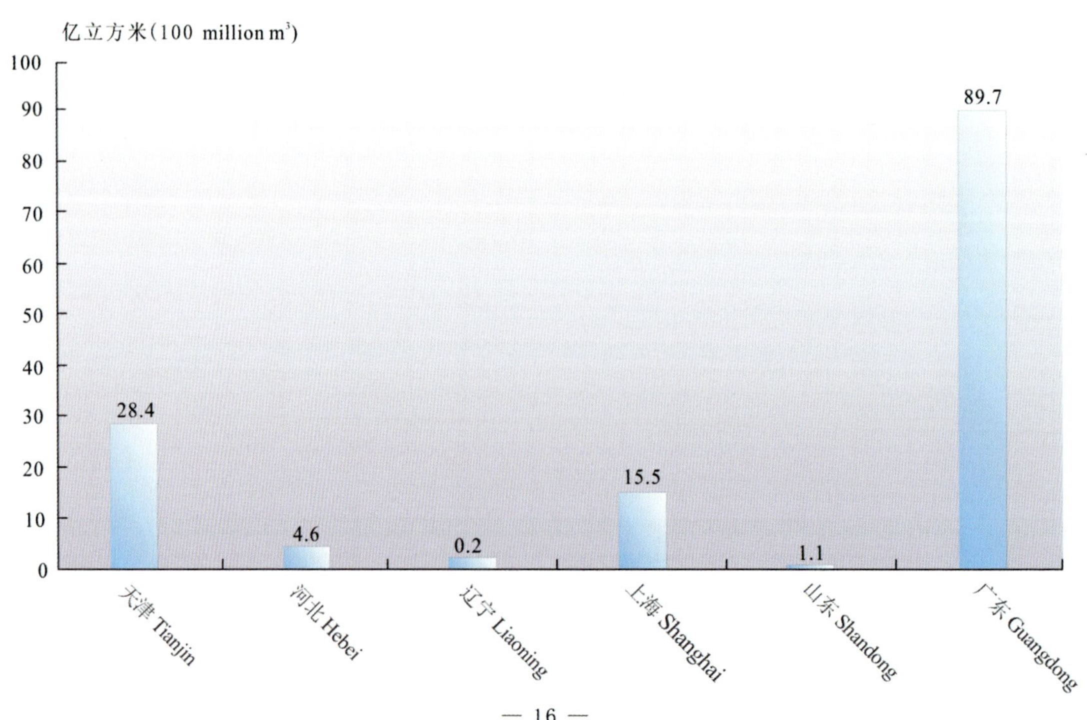

图 9 2017年海洋矿业产量

Production of Marine Mining Industry in 2017

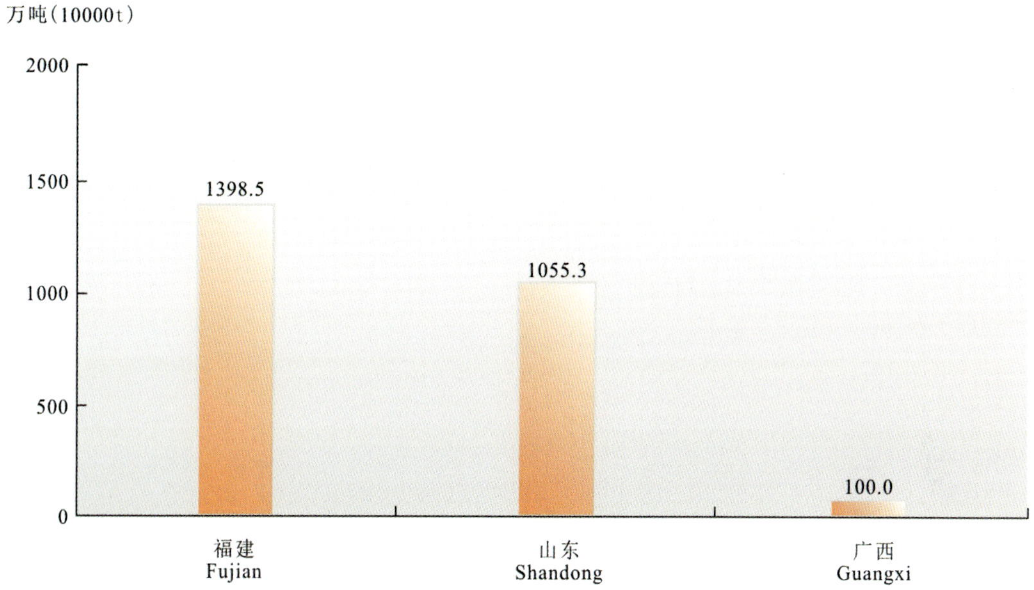

图 10 2017年沿海地区海盐产量

Sea Salt Production by Coastal Regions in 2017

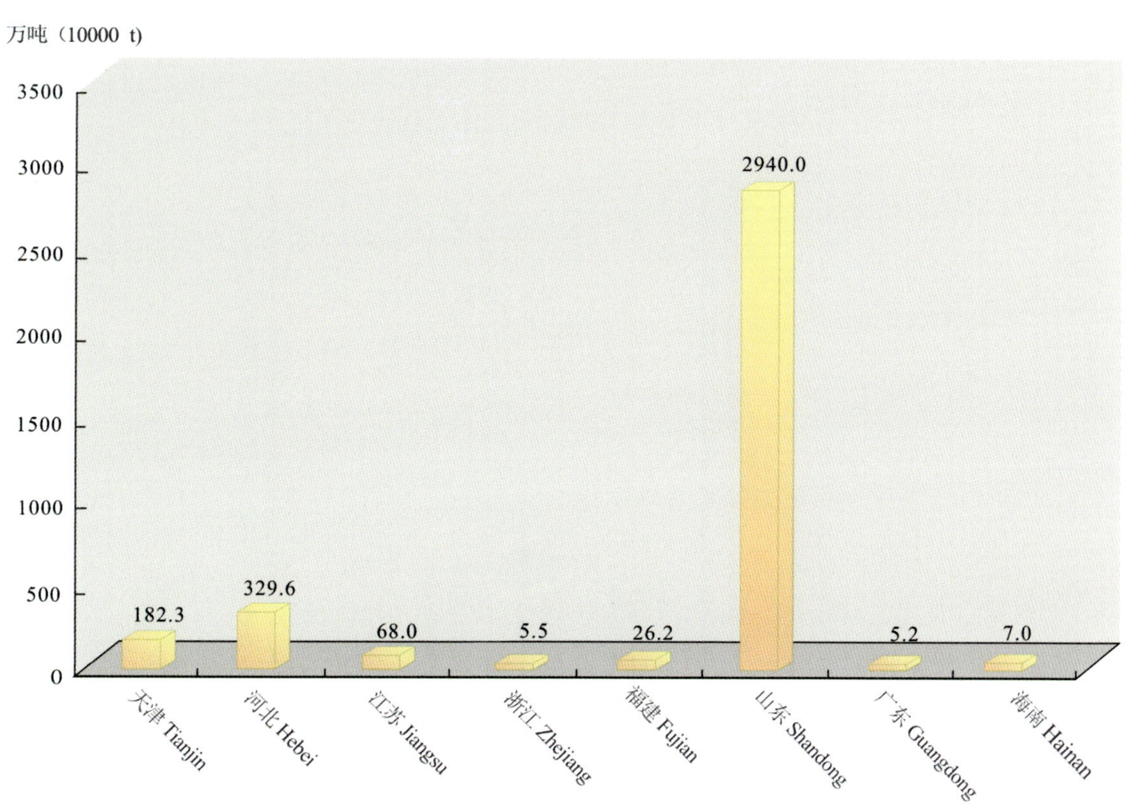

图 11 2017年沿海地区海洋造船完工量
Completed Quantity of Marine Shipbuilding by Coastal Regions in 2017

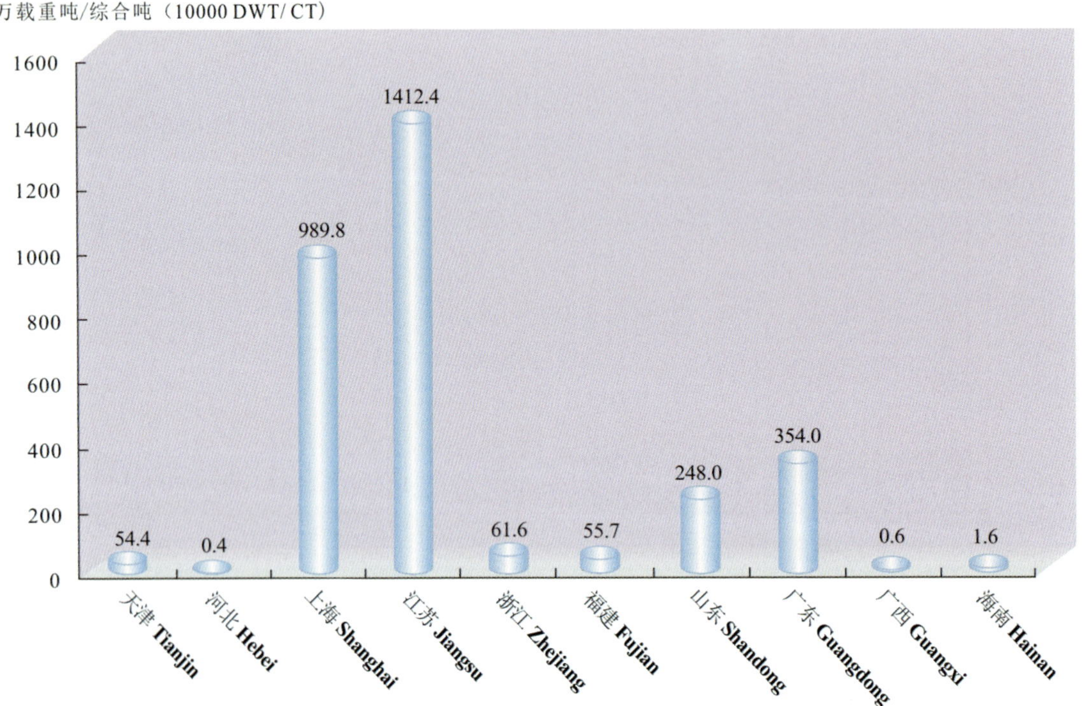

图 12 2017年沿海地区海洋货物周转量
Maritime Goods Turnover Volume by Coastal Regions in 2017

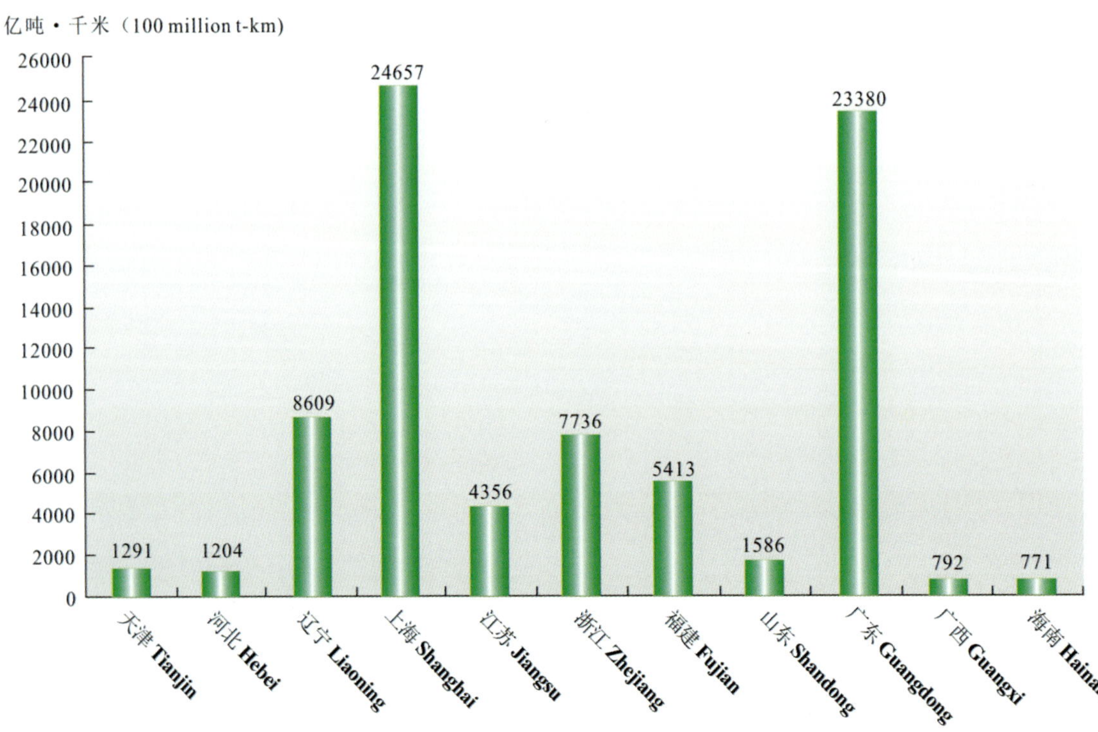

图 13 2017年沿海港口国际标准集装箱吞吐量

International Standardized Containers Handled by Coastal Seaports in 2017

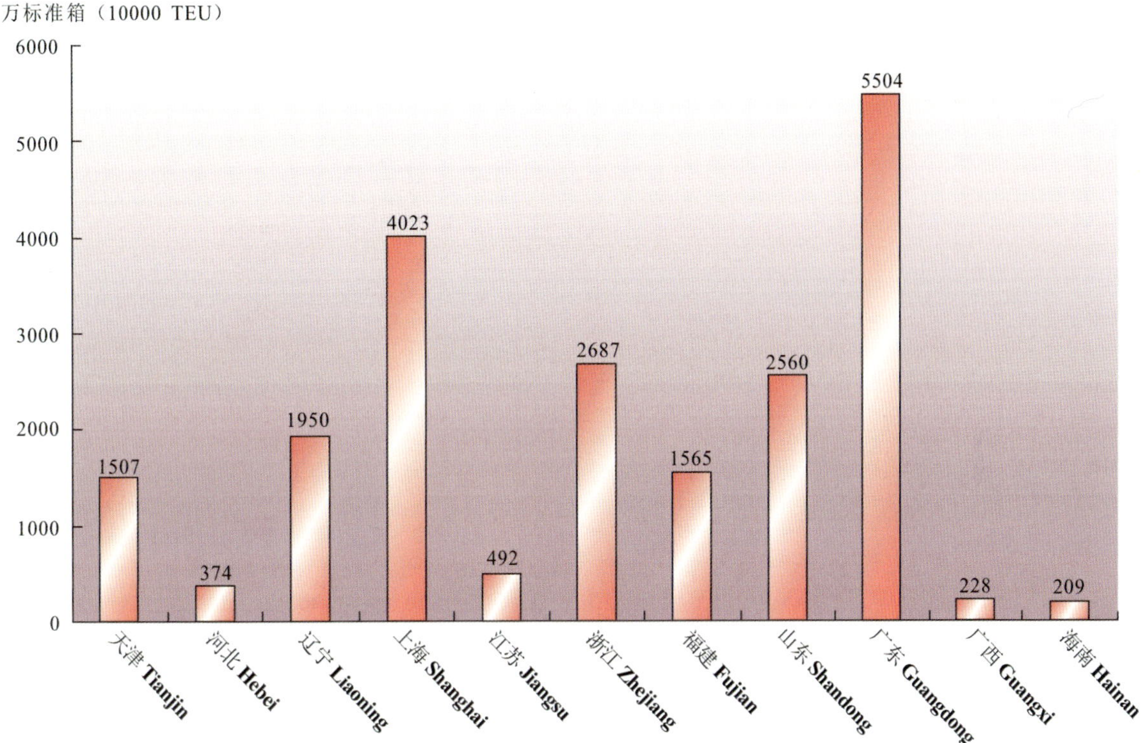

图 14 主要沿海城市接待入境过夜游客人数

Number of Oversea Visitor Arrivals in Major Coastal Cities

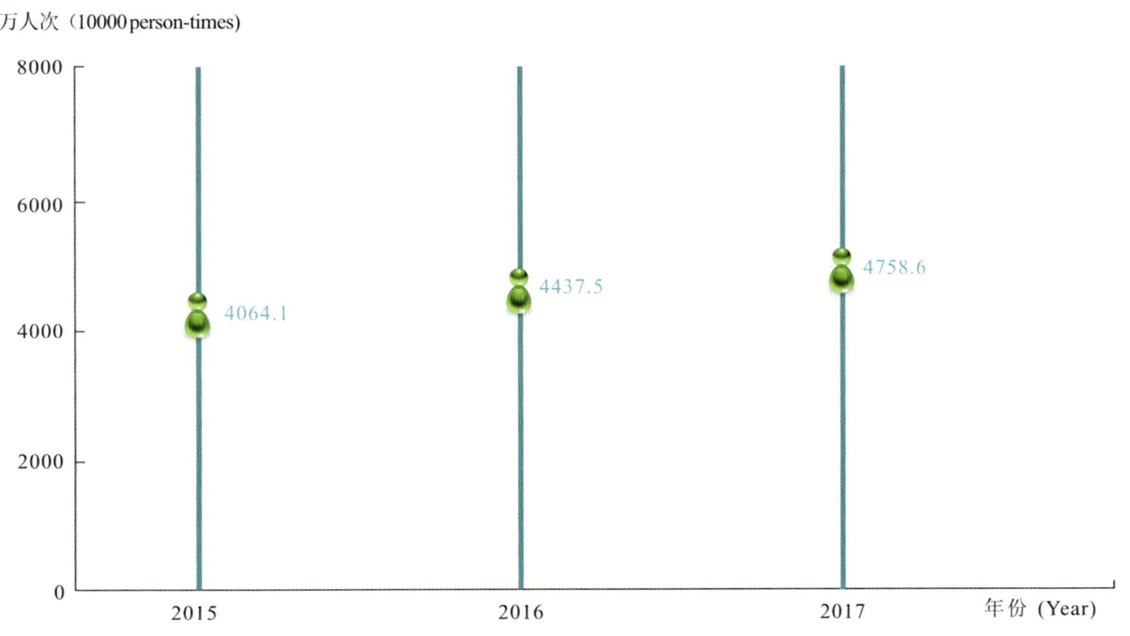

图 15 2017年环渤海、长江三角洲、珠江三角洲地区生产总值与海洋生产总值

GDP and GOP in the Round-the-Bohai Sea, Yangtze River Delta, and Zhujiang River Delta Regions in 2017

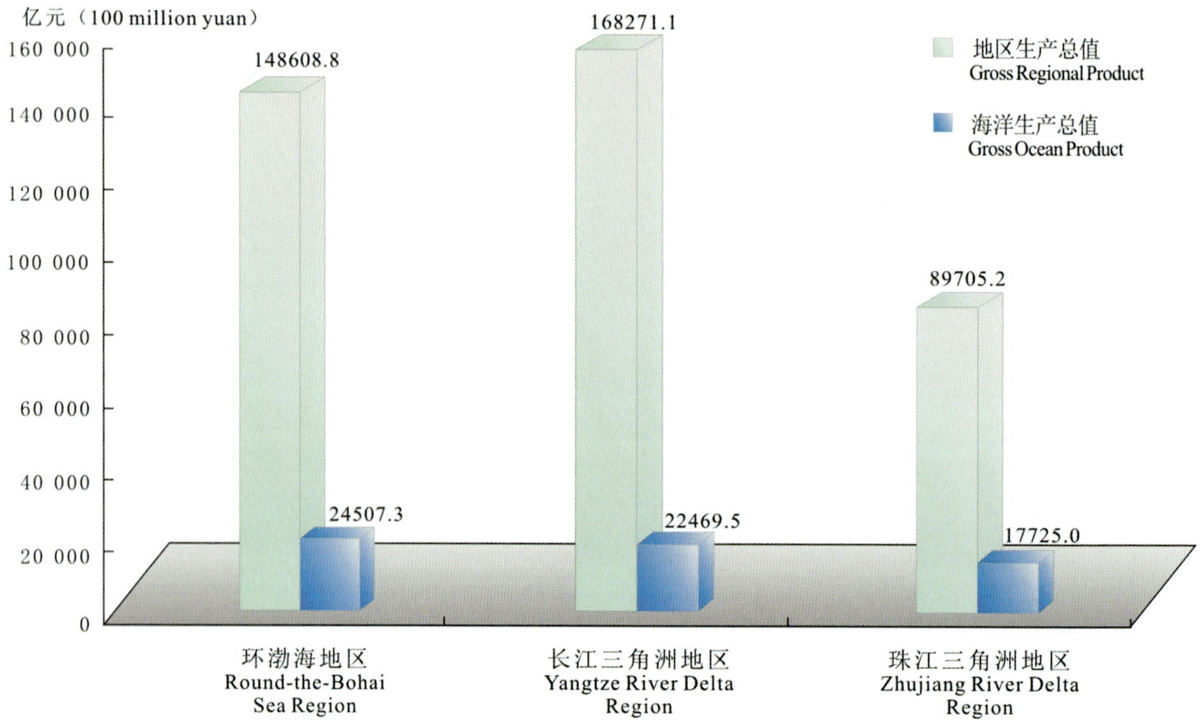

图 16 2017年环渤海、长江三角洲、珠江三角洲地区海洋生产总值占全国海洋生产总值比重

Proportion of the GOP of the Round-the-Bohai Sea, Yangtze River Delta, and Zhujiang River Delta Regions in the National GOP in 2017

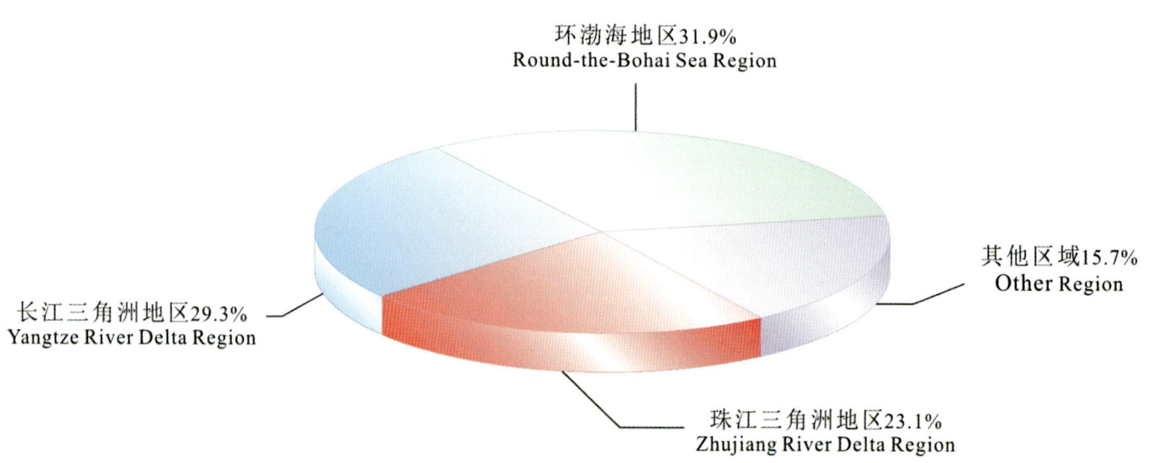

图 17 2017年沿海地区海洋生产总值

Gross Ocean Product by Coastal Regions in 2017

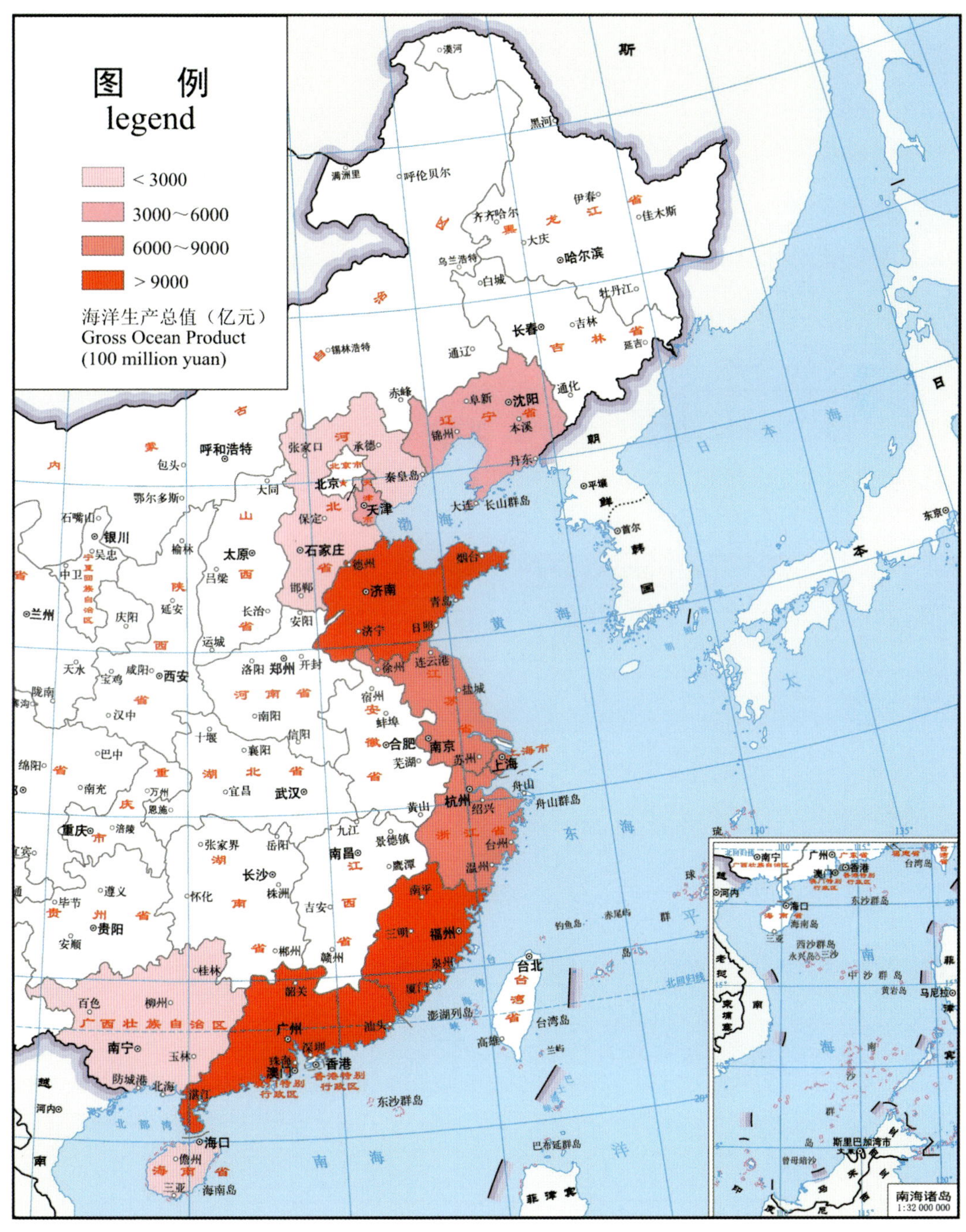

图 18 2017年沿海地区海洋经济贡献
Marine Economic Contributions by Coastal Regions in 2017

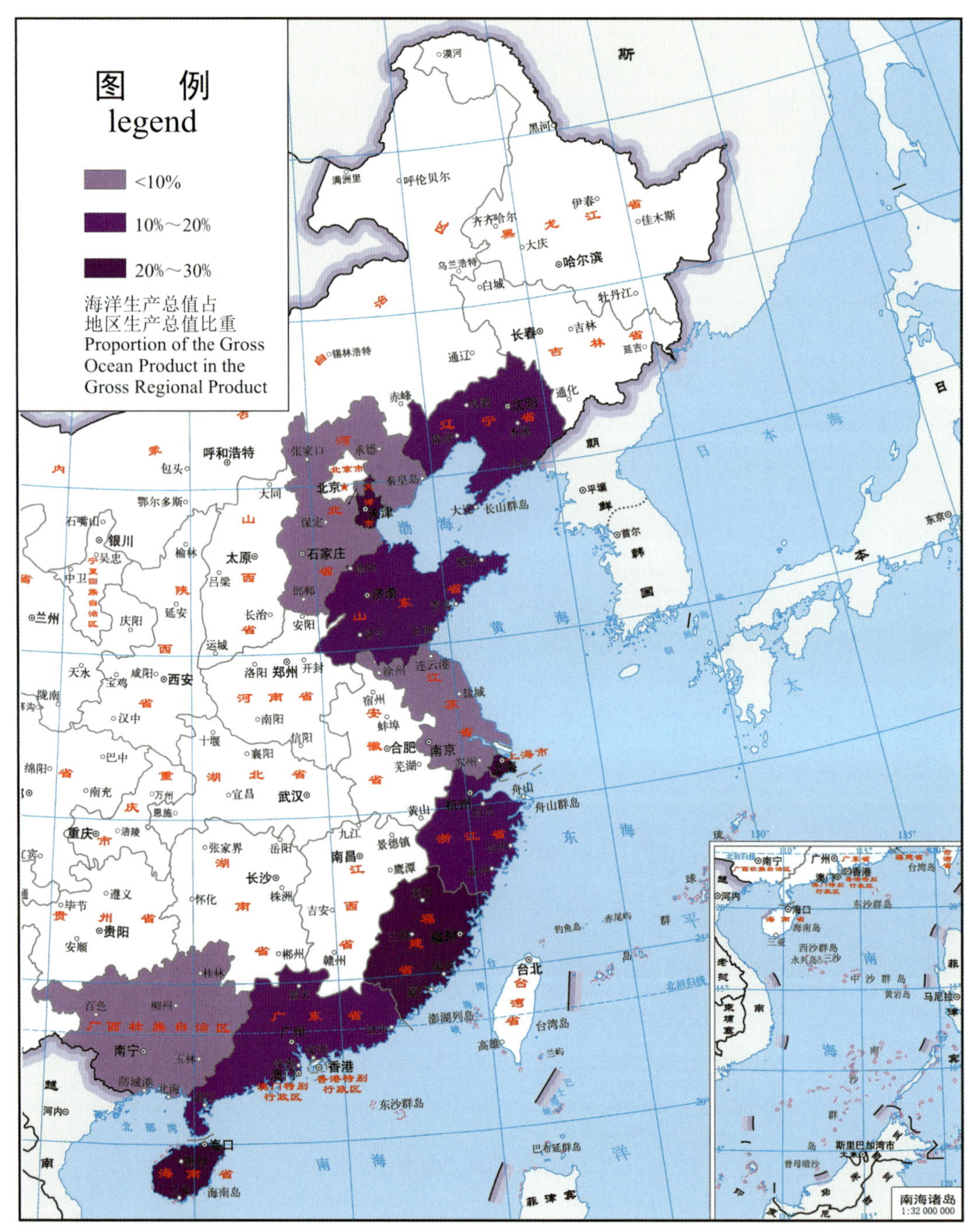

1

综 合 资 料

Integrated Data

1-1 沿海地区行政区划
Administrative Division of Coastal Regions

单位：个 (unit)

沿海地区 Coastal Region	沿海城市 Coastal City	沿海地带 Coastal County (District)			
		合　计 Total	县 County	县级市 County-level City	区 District
合　计 Total	**55**	**225**	**51**	**48**	**126**
天　津 Tianjin	1	1	0	0	1
河　北 Hebei	3	11	4	1	6
辽　宁 Liaoning	6	22	3	6	13
上　海 Shanghai	1	5	0	0	5
江　苏 Jiangsu	3	15	7	3	5
浙　江 Zhejiang	7	33	8	9	16
福　建 Fujian	6	33	11	7	15
山　东 Shandong	7	35	4	11	20
广　东 Guangdong	14	46	8	6	32
广　西 Guangxi	3	8	1	1	6
海　南 Hainan	4	16	5	4	7

注：沿海地带中未包括广东省的东莞、中山和海南的三沙、儋州。

Note: The coastal zone does not include Dongguan and Zhongshan of Guangdong Province, and Sansha and Danzhou of Hainan Province.

1-2 沿海行政区划一览表

Table of Administrative Division of Coastal Regions

沿海地区 Coastal Region	地区代码 Zip Code	沿海城市 Coastal City	地区代码 Zip Code	沿海地带 Coastal County (District)	地区代码 Zip Code
天 津 Tianjin	120000			滨海新区Binhai Xinqu	120116
河 北 Hebei	130000	唐山 Tangshan	130200	丰南区Fengnan Qu	130207
				曹妃甸区Caofeidian Qu	130209
				滦南县Luannan Xian	130224
				乐亭县Laoting Xian	130225
		秦皇岛 Qinhuangdao	130300	海港区Haigang Qu	130302
				山海关区Shanhaiguan Qu	130303
				北戴河区Beidaihe Qu	130304
				抚宁区Funing Qu	130306
				昌黎县Changli Xian	130322
		沧州 Cangzhou	130900	海兴县Haixing Xian	130924
				黄骅市Huanghua Shi	130983
辽 宁 Liaoning	210000	大连 Dalian	210200	中山区Zhongshan Qu	210202
				西岗区Xigang Qu	210203
				沙河口区Shahekou Qu	210204
				甘井子区Ganjingzi Qu	210211
				旅顺口区Lüshunkou Qu	210212
				金州区Jinzhou Qu	210213
				普兰店区Pulandian Qu	210214
				长海县Changhai Xian	210224
				瓦房店市Wafangdian Shi	210281
				庄河市Zhuanghe Shi	210283
		丹东 Dandong	210600	振兴区Zhenxing Qu	210603
				东港市Donggang Shi	210681
		锦州 Jinzhou	210700	凌海市Linghai Shi	210781
		营口 Yingkou	210800	鲅鱼圈区Bayuquan Qu	210804
				老边区Laobian Qu	210811
				盖州市Gaizhou Shi	210881
		盘锦 Panjin	211100	大洼区Dawa Qu	211104
				盘山县Panshan Xian	211122

1-2 续表1 continued

沿海地区 Coastal Region	地区代码 Zip Code	沿海城市 Coastal City	地区代码 Zip Code	沿海地带 Coastal County (District)	地区代码 Zip Code
		葫芦岛 Huludao	211400	连山区Lianshan Qu	211402
				龙港区Longgang Qu	211403
				绥中县Suizhong Xian	211421
				兴城市Xingcheng Shi	211481
上海 Shanghai	310000			宝山区Baoshan Qu	310113
				浦东新区Pudong Xinqu	310115
				金山区Jinshan Qu	310116
				奉贤区Fengxian Qu	310120
				崇明区Chongming Qu	310151
江苏 Jiangsu	320000	南通 Nantong	320600	通州区Tongzhou Qu	320612
				海安县Hai'an Xian	320621
				如东县Rudong Xian	320623
				启东市Qidong Shi	320681
				海门市Haimen Shi	320684
		连云港 Lianyungang	320700	连云区Lianyun Qu	320703
				赣榆区Ganyu Qu	320707
				灌云县Guanyun Xian	320723
				灌南县Guannan Xian	320724
		盐城 Yancheng	320900	亭湖区Tinghu Qu	320902
				大丰区Dafeng Qu	320904
				响水县Xiangshui Xian	320921
				滨海县Binhai Xian	320922
				射阳县Sheyang Xian	320924
				东台市Dongtai Shi	320981
浙江 Zhejiang	330000	杭州 Hangzhou	330100	滨江区Binjiang Qu	330108
				萧山区Xiaoshan Qu	330109
		宁波 Ningbo	330200	北仑区Beilun Qu	330206
				镇海区Zhenhai Qu	330211
				鄞州区Yinzhou Qu	330212
				奉化区Fenghua Qu	330213
				象山县Xiangshan Xian	330225
				宁海县Ninghai Xian	330226
				余姚市Yuyao Shi	330281
				慈溪市Cixi Shi	330282

1-2 续表2 continued

沿海地区 Coastal Region	地区代码 Zip Code	沿海城市 Coastal City	地区代码 Zip Code	沿海地带 Coastal County (District)	地区代码 Zip Code
		温州 Wenzhou	330300	鹿城区Lucheng Qu	330302
				龙湾区Longwan Qu	330303
				瓯海区Ouhai Qu	330304
				洞头区Dongtou Qu	330305
				平阳县Pingyang Xian	330326
				苍南县Cangnan Xian	330327
				瑞安市Rui'an Shi	330381
				乐清市Yueqing Shi	330382
		嘉兴 Jiaxing	330400	海盐县Haiyan Xian	330424
				海宁市Haining Shi	330481
				平湖市Pinghu Shi	330482
		绍兴 Shaoxing	330600	柯桥区Keqiao Qu	330603
				上虞区Shangyu Qu	330604
		舟山 Zhoushan	330900	定海区Dinghai Qu	330902
				普陀区Putuo Qu	330903
				岱山县Daishan Xian	330921
				嵊泗县Shengsi Xian	330922
		台州 Taizhou	331000	椒江区Jiaojiang Qu	331002
				路桥区Luqiao Qu	331004
				三门县Sanmen Xian	331022
				温岭市Wenling Shi	331081
				临海市Linhai Shi	331082
				玉环市Yuhuan Shi	331083
福 建 Fujian	350000	福州 Fuzhou	350100	马尾区Mawei Qu	350105
				长乐区Changle Qu	350112
				连江县Lianjiang Xian	350122
				罗源县Luoyuan Xian	350123
				平潭县Pingtan Xian	350128
				福清市Fuqing Shi	350181
		厦门 Xiamen	350200	思明区Siming Qu	350203
				海沧区Haicang Qu	350205

1-2 续表3 continued

沿海地区 Coastal Region	地区代码 Zip Code	沿海城市 Coastal City	地区代码 Zip Code	沿海地带 Coastal County (District)	地区代码 Zip Code
				湖里区Huli Qu	350206
				集美区Jimei Qu	350211
				同安区Tong'an Qu	350212
				翔安区Xiang'an Qu	350213
		莆田 Putian	350300	城厢区Chengxiang Qu	350302
				涵江区Hanjiang Qu	350303
				荔城区Licheng Qu	350304
				秀屿区Xiuyu Qu	350305
				仙游县Xianyou Xian	350322
		泉州 Quanzhou	350500	丰泽区Fengze Qu	350503
				泉港区Quangang Qu	350505
				惠安县Hui'an Xian	350521
				金门县Jinmen Xian	350527
				石狮市Shishi Shi	350581
				晋江市Jinjiang Shi	350582
				南安市Nan'an Shi	350583
		漳州 Zhangzhou	350600	云霄县Yunxiao Xian	350622
				漳浦县Zhangpu Xian	350623
				诏安县Zhao'an Xian	350624
				东山县Dongshan Xian	350626
				龙海市Longhai Shi	350681
		宁德 Ningde	350900	蕉城区Jiaocheng Qu	350902
				霞浦县Xiapu Xian	350921
				福安市Fu'an Shi	350981
				福鼎市Fuding Shi	350982
山 东 Shandong	370000	青岛 Qingdao	370200	市南区Shinan Qu	370202
				市北区Shibei Qu	370203
				黄岛区Huangdao Qu	370211
				崂山区Laoshan Qu	370212
				李沧区Licang Qu	370213
				城阳区Chengyang Qu	370214
				即墨区Jimo Qu	370215
				胶州市Jiaozhou Shi	370281

1-2 续表4 continued

沿海地区 Coastal Region	地区代码 Zip Code	沿海城市 Coastal City	地区代码 Zip Code	沿海地带 Coastal County (District)	地区代码 Zip Code
		东营 Dongying	370500	东营区Dongying Qu	370502
				河口区Hekou Qu	370503
				垦利区Kenli Qu	370505
				利津县Lijin Xian	370522
				广饶县Guangrao Xian	370523
		烟台 Yantai	370600	芝罘区Zhifu Qu	370602
				福山区Fushan Qu	370611
				牟平区Muping Qu	370612
				莱山区Laishan Qu	370613
				长岛县Changdao Xian	370634
				龙口市Longkou Shi	370681
				莱阳市Laiyang Shi	370682
				莱州市Laizhou Shi	370683
				蓬莱市Penglai Shi	370684
				招远市Zhaoyuan Shi	370685
				海阳市Haiyang Shi	370687
		潍坊 Weifang	370700	寒亭区Hanting Qu	370703
				寿光市Shouguang Shi	370783
				昌邑市Changyi Shi	370786
		威海 Weihai	371000	环翠区Huancui Qu	371002
				文登区Wendeng Qu	371003
				荣成市Rongcheng Shi	371082
				乳山市Rushan Shi	371083
		日照 Rizhao	371100	东港区Donggang Qu	371102
				岚山区Lanshan Qu	371103
		滨州 Binzhou	371600	沾化区Zhanhua Qu	371603
				无棣县Wudi Xian	371623
广东 Guangdong	440000	广州 Guangzhou	440100	黄埔区Huangpu Qu	440112
				番禺区Panyu Qu	440113
				南沙区Nansha Qu	440115
				增城区Zengcheng Qu	440118

1-2 续表5 continued

沿海地区 Coastal Region	地区代码 Zip Code	沿海城市 Coastal City	地区代码 Zip Code	沿海地带 Coastal County (District)	地区代码 Zip Code
		深圳 Shenzhen	440300	福田区Futian Qu	440304
				南山区Nanshan Qu	440305
				宝安区Bao'an Qu	440306
				龙岗区Longgang Qu	440307
				盐田区Yantian Qu	440308
		珠海 Zhuhai	440400	香洲区Xiangzhou Qu	440402
				斗门区Doumen Qu	440403
				金湾区Jinwan Qu	440404
		汕头 Shantou	440500	龙湖区Longhu Qu	440507
				金平区Jinping Qu	440511
				濠江区Haojiang Qu	440512
				潮阳区Chaoyang Qu	440513
				潮南区Chaonan Qu	440514
				澄海区Chenghai Qu	440515
				南澳县Nan'ao Xian	440523
		江门 Jiangmen	440700	蓬江区Pengjiang Qu	440703
				江海区Jianghai Qu	440704
				新会区Xinhui Qu	440705
				台山市Taishan Shi	440781
				恩平市Enping Shi	440785
		湛江 Zhanjiang	440800	赤坎区Chikan Qu	440802
				霞山区Xiashan Qu	440803
				坡头区Potou Qu	440804
				麻章区Mazhang Qu	440811
				遂溪县Suixi Xian	440823
				徐闻县Xuwen Xian	440825
				廉江市Lianjiang Shi	440881
				雷州市Leizhou Shi	440882
				吴川市Wuchuan Shi	440883
		茂名 Maoming	440900	电白区Dianbai Qu	440904
		惠州 Huizhou	441300	惠阳区Huiyang Qu	441303
				惠东县Huidong Xian	441323
		汕尾 Shanwei	441500	城　区Chengqu	441502
				海丰县Haifeng Xian	441521
				陆丰市Lufeng Shi	441581

1-2 续表6 continued

沿海地区 Coastal Region	地区代码 Zip Code	沿海城市 Coastal City	地区代码 Zip Code	沿海地带 Coastal County (District)	地区代码 Zip Code
		阳江 Yangjiang	441700	江城区Jiangcheng Qu	441702
				阳东区Yangdong Qu	441704
				阳西县Yangxi Xian	441721
		东莞 Dongguan	441900		
		中山 Zhongshan	442000		
		潮州 Chaozhou	445100	饶平县Raoping Xian	445122
		揭阳 Jieyang	445200	榕城区Rongcheng Qu	445202
				揭东区Jiedong Qu	445203
				惠来县Huilai Xian	445224
广 西 Guangxi	450000	北海 Beihai	450500	海城区Haicheng Qu	450502
				银海区Yinhai Qu	450503
				铁山港区Tieshangang Qu	450512
				合浦县Hepu Xian	450521
		防城港 Fangchenggang	450600	港口区Gangkou Qu	450602
				防城区Fangcheng Qu	450603
				东兴市Dongxing Shi	450681
		钦州 Qinzhou	450700	钦南区Qinnan Qu	450702
海 南 Hainan	460000	海口 Haikou	460100	秀英区Xiuying Qu	460105
				龙华区Longhua Qu	460106
				美兰区Meilan Qu	460108
		三亚 Sanya	460200	海棠区Haitang Qu	460202
				吉阳区Jiyang Qu	460203
				天涯区Tianya Qu	460204
				崖州区Yazhou Qu	460205
		三沙 Sansha	460300		
		儋州 Danzhou	460400		
		省直辖县 Counties Directly under the Hainan Province Government	469000	琼海市Qionghai Shi	469002
				文昌市Wenchang Shi	469005
				万宁市Wanning Shi	469006
				东方市Dongfang Shi	469007
				澄迈县Chengmai Xian	469023
				临高县Lingao Xian	469024
				昌江黎族自治县 Changjiang Lizu Zizhixian	469026
				乐东黎族自治县Ledong Lizu Zizhixian	469027
				陵水黎族自治县Lingshui Lizu Zizhixian	469028

1-3 海洋自然地理
Marine Physical Geography

指　　标	Item	指标值 Data
海洋平均深度　（米）	Average Depth of Sea (m)	961
海洋最大深度　（米）	Maximum Depth of Sea (m)	5559
岸线总长度　（千米）	Total Length of Coastline (km)	32000
大陆岸线长度	Length of Continental Coastline	18000
岛屿岸线长度	Length of Insular Coastline	14000
＞500 平方米岛屿数（个）	Number of Islands ＞500 m^2 each (unit)	7300
岛屿面积　（万平方千米）	Area of Islands (10 000 km^2)	8

注：海岛数据来源于《全国海岛保护规划》。

Note: Data on islands are derived from the *National Plan for Island Protection* .

1-4 海区海洋石油天然气储量
Offshore Oil and Natural Gas Reserves in the Sea Area

海 区 Sea Area	海洋石油（万吨） Offshore Oil (10000 t)		海洋天然气（亿立方米） Natural Gas (100 million m^3)	
	累计探明技术可采储量 Proven Technically Recoverable Reserves in the Aggregate	剩余技术可采储量 Surplus Technically Recoverable Reserves	累计探明技术可采储量 Proven Technically Recoverable Reserves in the Aggregate	剩余技术可采储量 Surplus Technically Recoverable Reserves
合 计 **Total**	**131367.2**	**63666.7**	**6777.4**	**5094.8**
渤 海 Bohai Sea	84592.1	49631.2	901.2	575.8
东 海 East China Sea	1437.7	890.7	1824.0	1685.3
南 海 South China Sea	45337.4	13144.8	4052.2	2833.7

注：数据来源于《2017年全国矿产资源储量通报》。
Note: The data come from the *Journal on the National Mineral Resources Reserves in 2017.*

1-5 沿海地区水资源情况
Water Resources by Coastal Regions

地 区 Region	水资源总量（亿立方米）Total Amount of Water Resources (100 million m^3)	地表水资源量 Surface Water Resources	地下水资源量 Groundwater Resources	地表水与地下水资源重复量 Duplicate Measurement of Surface Water and Groundwater	人均水资源量（立方米/人）Per Capita Water Resources (m^3/person)
全国总计 National Total	**28761.2**	**27746.3**	**8309.6**	**7294.7**	**2074.5**
天 津 Tianjin	13.0	8.8	5.5	1.3	83.4
河 北 Hebei	138.3	60.0	116.3	38.0	184.5
辽 宁 Liaoning	186.3	161.0	86.6	61.3	426.0
上 海 Shanghai	34.0	27.8	9.2	3.0	140.6
江 苏 Jiangsu	392.9	295.4	114.5	17.0	490.3
浙 江 Zhejiang	895.3	881.9	204.3	190.9	1592.1
福 建 Fujian	1055.6	1054.2	287.5	286.1	2711.9
山 东 Shandong	225.6	139.1	151.1	64.6	226.1
广 东 Guangdong	1786.6	1777.0	440.7	431.1	1611.9
广 西 Guangxi	2388.0	2386.0	446.6	444.6	4912.1
海 南 Hainan	383.9	380.5	96.8	93.4	4165.7

注：数据来源于《2018中国统计年鉴》。

Note: The data come from the *China Statistical Yearbook 2018.*

1-6 沿海地区湿地面积
Area of Wetlands by Coastal Regions

地 区 Region	湿地总面积 （千公顷） Total Area of Wetlands (1000 hm²)	#近海与海岸 Inshore and Coasts
全国总计 **National Total**	**53602.6**	**5795.9**
天 津 Tianjin	295.6	104.3
河 北 Hebei	941.9	231.9
辽 宁 Liaoning	1394.8	713.2
上 海 Shanghai	464.6	386.6
江 苏 Jiangsu	2822.8	1087.5
浙 江 Zhejiang	1110.1	692.5
福 建 Fujian	871.0	575.6
山 东 Shandong	1737.5	728.5
广 东 Guangdong	1753.4	815.1
广 西 Guangxi	754.3	259.0
海 南 Hainan	320.0	201.7

注：数据来源于《2018中国统计年鉴》。

Note: The data come from the *China Statistical Yearbook 2018.*

1-7 红树林各地类面积
Site Classification and Area of Sharpleaf Mangrove (*Rhizophora Apiculat*)

单位：公顷　　　　(hm²)

地　区 Region	红树林各地类总面积 Total Site Area of Sharpleaf Mangrove	现有面积 Established	未成林面积 Unestablished	宜林地面积 Suitable for Planting
全国总计 National Total	**82757.2**	**22024.9**	**1884.1**	**58848.2**
浙　江 Zhejiang	5452.3	20.6	236.1	5195.6
福　建 Fujian	13410.1	615.1	286.4	12508.6
广　东 Guangdong	32325.9	9084.0	981.3	22260.6
广　西 Guangxi	18029.2	8374.9	380.3	9274.0
海　南 Hainan	13539.7	3930.3		9609.4

注：数据来源于国家林业和草原局（原国家林业局）湿地调查。

Note: The data come from Wetland Survey of the State Forestry and Grassland Administration (originally the State Forestry Administration).

1-8 主要沿海城市气候基本情况
Climate of Major Coastal Cities

城 市 City	年平均气温（摄氏度）Annual Average Temperature (℃)	年平均相对湿度（%）Annual Average Relative Humidity (%)	全年降水量（毫米）Annual Precipitation (mm)	全年日照时数（小时）Annual Sunshine Hours (h)
天 津 Tianjin	14.2	55.0	512.9	2452.8
大 连 Dalian	12.4	58.0	521.7	2684.0
上 海 Shanghai	17.7	71.0	1388.8	1809.2
杭 州 Hangzhou	18.3	71.0	1442.0	1818.4
福 州 Fuzhou	21.2	73.0	1478.0	1632.4
青 岛 Qingdao	14.1	67.0	729.1	2314.0
广 州 Guangzhou	22.1	80.0	2067.4	1671.5
海 口 Haikou	24.6	82.0	2009.4	1994.0

注：数据来源于《2018中国统计年鉴》。

Note: The data come from the *China Statistical Yearbook 2018.*

主要统计指标解释

1. 沿海地区　是指有海岸线（大陆岸线和岛屿岸线）的地区，按行政区划分为沿海省、自治区、直辖市。

2. 沿海城市　是指有海岸线的直辖市和地级市（包括其下属的全部区、县和县级市）。

3. 沿海地带　是指有海岸线的县、县级市、区（包括直辖市和地级市的区）。

4. 海洋　是海和洋的统称。洋为地球表面上相连接的广大咸水水体的主体部分。海为地球表面相连接的广大咸水水体被陆地、岛礁、半岛包围或分隔的边缘部分。

5. 水资源总量　指评价区内降水形成的地表和地下产水总量，即地表产流量与降水入渗补给地下水量之和，不包括过境水量。

6. 地表水资源量　指评价区内河流、湖泊、冰川等地表水体中可以逐年更新的动态水量，即当地天然河川径流量。

7. 地下水资源量　指评价区内降水和地表水对饱水岩土层的补给量，包括降水入渗补给量和河道、湖库、渠系、渠灌田间等地表水体的入渗补给量。

8. 地表水与地下水资源重复量　指地表水和地下水相互转化的部分，即天然河川径流量中的地下水排泄量和地下水补给量中来源于地表水的入渗补给量。

9. 湿地　指天然或人工、长久或暂时性的沼泽地、泥炭地或水域地带，包括静止或流动、淡水、半咸水、咸水体，低潮时水深不超过6米的水域以及海岸地带地区的珊瑚滩和海草床、滩涂、红树林、河口、河流、淡水沼泽、沼泽森林、湖泊、盐沼及盐湖。

10. 红树林　指生长在热带、亚热带低能海岸潮间带上部，受周期性潮水浸淹，以红树植物为主体的常绿灌木或乔木组成的潮滩湿地木本生物群落。

11. 气温　指空气的温度，我国一般以摄氏度(℃)为单位表示。气象观测的温度表是放在离地面约1.5米处通风良好的百叶箱里测量的，因此，通常说的气温指的是离地面1.5米处百叶箱中的温度。其统计计算方法为：

月平均气温是将全月各日的平均气温相加，除以该月的天数而得；

年平均气温是将12个月的月平均气温累加后除以12而得。

12. 相对湿度　指空气中实际所含水蒸气密度和同温度下饱和水蒸气密度的百分比值。其统计方法与气温相同。

13. 降水量　指从天空降落到地面的液态或固态(经融化后)水，未经蒸发、渗透、流失而在地面上积聚的深度。其统计计算方法为：

月降水量是将全月各日的降水量累加而得；

年降水量是将12个月的月降水量累加而得。

14. 日照时数　指太阳实际照射地面的时间。其统计方法与降水量相同。

Explanatory Notes on Main Statistical Indicators

1. Coastal Region refers to the regions with coastlines (continental and island coastlines), which are divided into the coastal provinces, autonomous regions and municipalities directly under the Central Government according to the administrative zoning.

2. Coastal City refers to the municipalities directly under the Central Government and the prefecture-level cities (including all the districts, counties and county-level cities under them).

3. Coastal County (District), i.e., the coastal region in a narrow sense, refers to the counties, county-level cities and districts with coastlines (including the districts under the municipalities directly under the Central Government and the prefecture-level districts).

4. Ocean is the general name for sea and ocean. Ocean refers to the main body of large salt water connected with the earth surface. Sea refers to the edge areas of the salt water on the earth surface that are compartmentalized or surrounded by land, island, reef or peninsula.

5. Total Water Resources refers to total volume of water resources measured as run-off for surface water from rainfall and recharge for groundwater in a given area, excluding transit water.

6. Surface Water Resources refers to total renewable resources which exist in rivers, lakes, glaciers and other collectors from rainfall and are measured as run-off of rivers.

7. Groundwater Resources refers to replenishment of aquifers with rainfall and surface water.

8. Duplicated Measurement between Surface Water and Groundwater refers to the exchange between surface water and groundwater, i.e. run-off of rivers includes some depletion into groundwater while groundwater includes some replenishment from surface water.

9. Wetland refers to marshland and peat bog, whether natural or man-made, permanent or temporary; water covered areas, whether stagnant or flowing, with fresh or brackish-fresh or salty water that is less than 6 meters deep at low tide; as well as coral beach, seagrass bed, tidal flat, mangrove, river outlet, rivers, freshwater marshland, marshland forests, lakes, salty bog and salt lakes along the coastal areas.

10. Mangrove refers to evergreen woody plants or plant communities in tropical or sub-tropical zones which live between the sea and the land in areas which are inundated by tides.

11. Temperature refers to the air temperature. China uses centigrade as the unit. The thermometry used for weather observation is put in a breezy shutter, which is 1.5 meters high from the ground. Therefore, the commonly used temperature refers to the temperature in the breezy shutter 1.5 meters above the ground. The calculation method is as follows:

Monthly average temperature is the summation of average daily temperature of one month divided by the actual days of that particular month.

Annual average temperature is the summation of monthly averages of a year divided by 12 months.

12. Relative Humidity refers to the ratio of actual water vapour pressure to the saturated water vapour density under the current temperature. The calculation method is the same as that for the temperature.

13. Volume of Precipitation refers to the deepness of liquid state or solid state (thawed) water falling from the sky to the ground that has not evaporated, infiltrated or run off. The calculation method is as follows:

Monthly precipitation is the summation of daily precipitation of a month.

Annual precipitation is the summation of 12 months precipitation of a year.

14. Sunshine Hours refers to the actual hours of sun irradiating the earth. The calculation method is the same as that for the precipitation.

2

海洋经济核算

Marine Economic Accounting

2-1 全国海洋生产总值
National Gross Ocean Product

年 份 Year	海洋生产总值（亿元） Gross Ocean Product (100 million yuan)				海洋生产总值占国内生产总值比重（%） Proportion of the Gross Ocean Product in GDP (%)	海洋生产总值增长速度（按可比价计算）（%） Growth Rate of the Gross Ocean Product (%)
		第一产业 Primary Industry	第二产业 Secondary Industry	第三产业 Tertiary Industry		
2001	9518.4	646.3	4152.1	4720.1	8.59	
2002	11270.5	730.0	4866.2	5674.3	9.26	19.8
2003	11952.3	766.2	5367.6	5818.5	8.70	4.2
2004	14662.0	851.0	6662.8	7148.2	9.06	16.9
2005	17655.6	1008.9	8046.9	8599.8	9.43	16.3
2006	21592.4	1228.8	10217.8	10145.7	9.84	18.0
2007	25618.7	1395.4	12011.0	12212.3	9.49	14.8
2008	29718.0	1694.3	13735.3	14288.4	9.31	9.9
2009	32161.9	1857.7	14926.5	15377.6	9.23	8.8
2010	39619.2	2008.0	18919.6	18691.6	9.61	15.3
2011	45580.4	2381.9	21667.6	21530.8	9.34	10.0
2012	50172.9	2670.6	23450.2	24052.1	9.32	8.1
2013	54718.3	3037.7	24608.9	27071.7	9.23	7.8
2014	60699.1	3109.5	26660.0	30929.6	9.47	7.9
2015	65534.4	3327.7	27671.9	34534.8	9.55	7.0
2016	69693.7	3570.9	27666.6	38456.2	9.42	6.7
2017	76749.0	3628.1	28951.9	44169.0	9.35	6.9

2-2 全国海洋生产总值构成
Composition of National Gross Ocean Product

单位：% (%)

年 份 Year	第一产业 Primary Industry	第二产业 Secondary Industry	第三产业 Tertiary Industry
2001	6.8	43.6	49.6
2002	6.5	43.2	50.3
2003	6.4	44.9	48.7
2004	5.8	45.4	48.8
2005	5.7	45.6	48.7
2006	5.7	47.3	47.0
2007	5.4	46.9	47.7
2008	5.7	46.2	48.1
2009	5.8	46.4	47.8
2010	5.1	47.8	47.2
2011	5.2	47.5	47.2
2012	5.3	46.7	47.9
2013	5.6	45.0	49.5
2014	5.1	43.9	51.0
2015	5.1	42.2	52.7
2016	5.1	39.7	55.2
2017	4.7	37.7	57.5

2-3 海洋及相关产业增加值
Added Values of Marine and Marine-Related Industries

单位：亿元 (100 million yuan)

年　份 Year	合 计 Total	海洋产业 Marine Industry	主要海洋产业 Major Marine Industry	海洋科研教育管理服务业 Marine Scientific Research, Education, Management and Service	海洋相关产业 Marine-related Industries
2001	9518.4	5733.6	3856.6	1877.0	3784.8
2002	11270.5	6787.3	4696.8	2090.5	4483.2
2003	11952.3	7137.7	4754.4	2383.3	4814.6
2004	14662.0	8710.1	5827.7	2882.5	5951.9
2005	17655.6	10539.0	7188.0	3350.9	7116.6
2006	21592.4	12696.7	8790.4	3906.4	8895.6
2007	25618.7	15070.6	10478.3	4592.3	10548.0
2008	29718.0	17591.2	12176.0	5415.2	12126.8
2009	32161.9	18769.4	12768.4	6001.0	13392.5
2010	39619.2	22886.4	16187.8	6698.5	16732.8
2011	45580.4	26517.6	18865.2	7652.4	19062.8
2012	50172.9	29404.6	20829.9	8574.8	20768.2
2013	54718.3	32658.7	22462.3	10196.4	22059.6
2014	60699.1	36364.9	25303.4	11061.5	24334.1
2015	65534.4	39554.9	26838.8	12716.0	25979.5
2016	69693.7	43013.0	28391.9	14621.1	26680.6
2017	76749.0	48327.3	31122.5	17204.8	28421.8

2-4 海洋及相关产业增加值构成
Composition of the Added Values of Marine and Marine-Related Industries

单位：% (%)

年　份 Year	合 计 Total	海洋产业 Marine Industry	主要海洋产业 Major Marine Industry	海洋科研教育管理服务业 Marine Scientific Research, Education, Management and Service	海洋相关产业 Marine-related Industries
2001	100.0	60.2	40.5	19.7	39.8
2002	100.0	60.2	41.7	18.5	39.8
2003	100.0	59.7	39.8	19.9	40.3
2004	100.0	59.4	39.7	19.7	40.6
2005	100.0	59.7	40.7	19.0	40.3
2006	100.0	58.8	40.7	18.1	41.2
2007	100.0	58.8	40.9	17.9	41.2
2008	100.0	59.2	41.0	18.2	40.8
2009	100.0	58.4	39.7	18.7	41.6
2010	100.0	57.8	40.9	16.9	42.2
2011	100.0	58.2	41.4	16.8	41.8
2012	100.0	58.6	41.5	17.1	41.4
2013	100.0	59.7	41.1	18.6	40.3
2014	100.0	59.9	41.7	18.2	40.1
2015	100.0	60.4	41.0	19.4	39.6
2016	100.0	61.7	40.7	21.0	38.3
2017	100.0	63.0	40.6	22.4	37.0

2-5 全国主要海洋产业增加值（2017年）
Added Values of National Major Marine Industries (2017)

主要海洋产业 Major Marine Industry	增加值 （亿元） Added Value (100 million yuan)	比上年增长（%） （按可比价计算） Percentage of Increase over Last Year (%) (at comparable price)
合计 Total	**31122.5**	**7.4**
海洋渔业 Marine Fishery Industry	4700.7	-2.8
海洋油气业 Offshore Oil and Gas Industry	1145.2	-0.4
海洋矿业 Marine Mining Industry	65.2	-7.5
海洋盐业 Sea Salt Industry	42.3	-7.5
海洋船舶工业 Marine Shipbuilding Industry	1091.5	-5.7
海洋化工业 Marine Chemical Industry	1021.0	-3.0
海洋生物医药业 Marine Biomedicine Industry	389.1	12.3
海洋工程建筑业 Marine Engineering Construction Industry	1846.4	-1.3
海洋电力业 Marine Electric Power Industry	151.7	18.8
海水利用业 Seawater Utilization Industry	15.8	15.3
海洋交通运输业 Marine Communications and Transport Industry	6081.0	5.5
滨海旅游业 Coastal Tourism	14572.5	16.0

2-6 海洋渔业增加值
Added Value of Marine Fishery Industry

单位：亿元 (100 million yuan)

年 份 Year	增加值 Added Value
2001	966.0
2002	1091.2
2003	1145.0
2004	1271.2
2005	1507.6
2006	1672.0
2007	1906.0
2008	2228.6
2009	2440.8
2010	2851.6
2011	3202.9
2012	3560.5
2013	3997.6
2014	4126.6
2015	4317.4
2016	4615.4
2017	4700.7

2-7 海洋油气业增加值
Added Value of Offshore Oil and Gas Industry

单位：亿元 (100 million yuan)

年 份 Year	增加值 Added Value
2001	176.8
2002	181.8
2003	257.0
2004	345.1
2005	528.2
2006	668.9
2007	666.9
2008	1020.5
2009	614.1
2010	1302.2
2011	1719.7
2012	1718.7
2013	1666.6
2014	1530.4
2015	981.9
2016	868.8
2017	1145.2

2-8 海洋矿业增加值
Added Value of Marine Mining Industry

单位：亿元 (100 million yuan)

年　份 Year	增加值 Added Value
2001	1.0
2002	1.9
2003	3.1
2004	7.9
2005	8.3
2006	13.4
2007	16.3
2008	35.2
2009	41.6
2010	45.2
2011	53.3
2012	45.1
2013	54.0
2014	59.6
2015	63.9
2016	67.3
2017	65.2

注：自2008年起部分地区统计矿种增加。

Note: The data from 2008 include added kinds of minerals in some regions.

2-9 海洋盐业增加值
Added Value of Sea Salt Industry

单位：亿元 (100 million yuan)

年　份 Year	增加值 Added Value
2001	32.6
2002	34.2
2003	28.4
2004	39.0
2005	39.1
2006	37.1
2007	39.9
2008	43.6
2009	43.6
2010	65.5
2011	76.8
2012	60.1
2013	63.2
2014	68.3
2015	41.0
2016	38.9
2017	42.3

2-10 海洋船舶工业增加值
Added Value of Marine Shipbuilding Industry

单位：亿元 (100 million yuan)

年　份 Year	增加值 Added Value
2001	109.3
2002	117.4
2003	152.8
2004	204.1
2005	275.5
2006	339.5
2007	524.9
2008	742.6
2009	986.5
2010	1215.6
2011	1352.0
2012	1291.3
2013	1213.2
2014	1395.5
2015	1445.7
2016	1492.4
2017	1091.5

2-11 海洋化工业增加值
Added Value of Marine Chemical Industry

单位：亿元 (100 million yuan)

年　份 Year	增加值 Added Value
2001	64.7
2002	77.1
2003	96.3
2004	151.5
2005	153.3
2006	440.4
2007	506.6
2008	416.8
2009	465.3
2010	613.8
2011	695.9
2012	843.0
2013	813.9
2014	920.0
2015	964.2
2016	961.8
2017	1021.0

注：自2006年起部分地区统计产品品种增加。

Note: The data from 2006 include added kinds of statistical products in some regions.

2-12 海洋生物医药业增加值
Added Value of Marine Biomedicine Industry

单位：亿元 (100 million yuan)

年　份 Year	增加值 Added Value
2001	5.7
2002	13.2
2003	16.5
2004	19.0
2005	28.6
2006	34.8
2007	45.4
2008	56.6
2009	52.1
2010	83.8
2011	150.8
2012	184.7
2013	238.7
2014	258.1
2015	295.7
2016	341.3
2017	389.1

2-13 海洋工程建筑业增加值
Added Value of Marine Engineering Construction Industry

单位：亿元 (100 million yuan)

年　份 Year	增加值 Added Value
2001	109.2
2002	145.4
2003	192.6
2004	231.8
2005	257.2
2006	423.7
2007	499.7
2008	347.8
2009	672.3
2010	874.2
2011	1086.8
2012	1353.8
2013	1595.5
2014	1735.0
2015	2073.5
2016	1731.3
2017	1846.4

2-14 海洋电力业增加值
Added Value of Marine Electric Power Industry

单位：亿元 (100 million yuan)

年 份 Year	增加值 Added Value
2001	1.8
2002	2.2
2003	2.8
2004	3.1
2005	3.5
2006	4.4
2007	5.1
2008	11.3
2009	20.8
2010	38.1
2011	59.2
2012	77.3
2013	91.5
2014	107.7
2015	120.1
2016	128.5
2017	151.7

2-15 海水利用业增加值
Added Value of Seawater Utilization Industry

单位：亿元 (100 million yuan)

年 份 Year	增加值 Added Value
2001	1.1
2002	1.3
2003	1.7
2004	2.4
2005	3.0
2006	5.2
2007	6.2
2008	7.4
2009	7.8
2010	8.9
2011	10.4
2012	11.1
2013	11.9
2014	12.7
2015	13.7
2016	13.7
2017	15.8

2-16 海洋交通运输业增加值
Added Value of Marine Communications and Transport Industry

单位：亿元 (100 million yuan)

年　份 Year	增加值 Added Value
2001	1316.4
2002	1507.4
2003	1752.5
2004	2030.7
2005	2373.3
2006	2531.4
2007	3035.6
2008	3499.3
2009	3146.6
2010	3785.8
2011	4217.5
2012	4752.6
2013	4876.5
2014	5336.9
2015	5641.1
2016	5699.8
2017	6081.0

2-17 滨海旅游业增加值
Added Value of Coastal Tourism

单位：亿元 (100 million yuan)

年　份 Year	增加值 Added Value
2001	1072.0
2002	1523.7
2003	1105.8
2004	1522.0
2005	2010.6
2006	2619.6
2007	3225.8
2008	3766.4
2009	4277.1
2010	5303.1
2011	6239.9
2012	6931.8
2013	7839.7
2014	9752.8
2015	10880.6
2016	12432.8
2017	14572.5

2-18 沿海地区海洋生产总值（2017年）
Gross Ocean Product by Coastal Regions (2017)

地 区 Region	海洋生产总值（亿元） Gross Ocean Product (100 million yuan)				海洋生产总值占地区生产总值比重（%） Proportion of the Gross Ocean Product in the Gross Regional Product (%)
		第一产业 Primary Industry	第二产业 Secondary Industry	第三产业 Tertiary Industry	
合 计 Total	**76749.0**	**3628.1**	**28951.9**	**44169.0**	**16.6**
天 津 Tianjin	4646.6	10.5	2155.2	2480.9	25.1
河 北 Hebei	2385.5	86.0	827.2	1472.3	7.0
辽 宁 Liaoning	3284.1	449.6	1045.0	1789.5	14.0
上 海 Shanghai	8494.7	4.9	2854.9	5634.9	27.7
江 苏 Jiangsu	6933.4	440.7	3163.7	3329.0	8.1
浙 江 Zhejiang	7041.4	517.7	2203.6	4320.1	13.6
福 建 Fujian	9384.0	598.1	3179.9	5606.0	29.2
山 东 Shandong	14191.1	720.0	6046.8	7424.4	19.5
广 东 Guangdong	17725.0	315.2	6776.0	10633.8	19.8
广 西 Guangxi	1377.0	219.4	462.0	695.6	7.4
海 南 Hainan	1286.3	266.1	237.7	782.5	28.8

2-19 沿海地区海洋生产总值构成（2017年）
Composition of Gross Ocean Product by Coastal Regions (2017)

单位：% (%)

地 区 Region	海洋生产总值 Gross Ocean Product			
		第一产业 Primary Industry	第二产业 Secondary Industry	第三产业 Tertiary Industry
合 计 Total	**100.0**	**4.7**	**37.7**	**57.5**
天 津 Tianjin	100.0	0.2	46.4	53.4
河 北 Hebei	100.0	3.6	34.7	61.7
辽 宁 Liaoning	100.0	13.7	31.8	54.5
上 海 Shanghai	100.0	0.1	33.6	66.3
江 苏 Jiangsu	100.0	6.4	45.6	48.0
浙 江 Zhejiang	100.0	7.4	31.3	61.4
福 建 Fujian	100.0	6.4	33.9	59.7
山 东 Shandong	100.0	5.1	42.6	52.3
广 东 Guangdong	100.0	1.8	38.2	60.0
广 西 Guangxi	100.0	15.9	33.5	50.5
海 南 Hainan	100.0	20.7	18.5	60.8

2-20 沿海地区海洋及相关产业增加值（2017年）
Added Values of Marine and Marine-Related Industries by Coastal Regions (2017)

单位：亿元 (100 million yuan)

地 区 Region	合 计 Total	海洋产业 Marine Industry	主要海洋产业 Major Marine Industry	海洋科研教育管理服务业 Marine Scientific Research, Education, Management and Service	海洋相关产业 Marine-related Industries
合 计 Total	**76749.0**	**48327.3**	**31122.5**	**17204.8**	**28421.8**
天 津 Tianjin	4646.6	2877.3	2464.7	412.6	1769.4
河 北 Hebei	2385.5	1533.3	1403.1	130.2	852.2
辽 宁 Liaoning	3284.1	2203.8	1630.7	573.1	1080.3
上 海 Shanghai	8494.7	5297.2	2686.1	2611.1	3197.5
江 苏 Jiangsu	6933.4	4137.7	2673.0	1464.7	2795.6
浙 江 Zhejiang	7041.4	4595.6	2798.3	1797.3	2445.8
福 建 Fujian	9384.0	5527.9	4223.3	1304.6	3856.2
山 东 Shandong	14191.1	8695.4	5832.0	2863.4	5495.8
广 东 Guangdong	17725.0	11671.0	6064.8	5606.2	6054.0
广 西 Guangxi	1377.0	876.8	733.9	142.9	500.2
海 南 Hainan	1286.3	911.5	612.6	298.9	374.9

2-21 沿海地区海洋及相关产业增加值构成（2017年）
Composition of the Added Values of Marine and Marine-Related Industries by Coastal Regions (2017)

单位：%　　　　(%)

地　区 Region	合　计 Total	海洋产业 Marine Industry	主要海洋产业 Major Marine Industry	海洋科研教育管理服务业 Marine Scientific Research, Education, Management and Service	海洋相关产业 Marine-related Industries
合　计 Total	**100.0**	**63.0**	**40.6**	**22.4**	**37.0**
天　津 Tianjin	100.0	61.9	53.0	8.9	38.1
河　北 Hebei	100.0	64.3	58.8	5.5	35.7
辽　宁 Liaoning	100.0	67.1	49.7	17.4	32.9
上　海 Shanghai	100.0	62.4	31.6	30.7	37.6
江　苏 Jiangsu	100.0	59.7	38.6	21.1	40.3
浙　江 Zhejiang	100.0	65.3	39.7	25.5	34.7
福　建 Fujian	100.0	58.9	45.0	13.9	41.1
山　东 Shandong	100.0	61.3	41.1	20.2	38.7
广　东 Guangdong	100.0	65.8	34.2	31.6	34.2
广　西 Guangxi	100.0	63.7	53.3	10.4	36.3
海　南 Hainan	100.0	70.9	47.6	23.2	29.1

主要统计指标解释

1. 海洋经济 是开发、利用和保护海洋的各类产业活动以及与之相关联活动的总和。

2. 海洋生产总值 是海洋经济生产总值的简称，指按市场价格计算的沿海地区常住单位在一定时期内海洋经济活动的最终成果，是海洋产业和海洋相关产业增加值之和。

3. 海洋产业 是开发、利用和保护海洋所进行的生产和服务活动，包括海洋渔业、海洋油气业、海洋矿业、海洋盐业、海洋化工业、海洋生物医药业、海洋电力业、海水利用业、海洋船舶工业、海洋工程建筑业、海洋交通运输业、滨海旅游业等主要海洋产业以及海洋科研教育管理服务业。

4. 海洋科研教育管理服务业 是开发、利用和保护海洋过程中所进行的科研、教育、管理及服务等活动，包括海洋信息服务业、海洋环境监测预报服务、海洋保险与社会保障业、海洋科学研究、海洋技术服务业、海洋地质勘查业、海洋环境保护业、海洋教育、海洋管理、海洋社会团体与国际组织等。

5. 海洋相关产业 是指以各种投入产出为联系纽带，与主要海洋产业构成技术经济联系的上下游产业，涉及海洋农林业、海洋设备制造业、涉海产品及材料制造业、涉海建筑与安装业、海洋批发与零售业、涉海服务业等。

6. 海洋三次产业 我国的海洋三次产业划分如下：

海洋第一产业：是指海洋渔业中的海洋水产品、海洋渔业服务业，以及海洋相关产业中属于第一产业范畴的部门。

海洋第二产业：是指海洋渔业中海洋水产品加工、海洋油气业、海洋矿业、海洋盐业、海洋化工业、海洋生物医药业、海洋电力业、海水利用业、海洋船舶工业、海洋工程建筑业，以及海洋相关产业中属于第二产业范畴的部门。

海洋第三产业：是指除海洋第一、第二产业以外的其他行业。第三产业包括海洋交通运输业、滨海旅游业、海洋科研教育管理服务业，以及海洋相关产业中属于第三产业范畴的部门。

7. 海洋渔业 包括海水养殖、海洋捕捞、海洋渔业服务业和海洋水产品加工等活动。

8. 海洋油气业 是指在海洋中勘探、开采、输送、加工原油和天然气的生产活动。

9. 海洋矿业 包括海滨砂矿、海滨土砂石与煤矿及深海矿物等的采选活动。

10. 海洋盐业 是指利用海水生产以氯化钠为主要成分的盐产品的活动，包括采盐和盐加工。

11. 海洋船舶工业 是指以金属或非金属为主要材料，制造海洋船舶、海上固定及浮动装置的活动，以及对海洋船舶的修理及拆卸活动。

12. 海洋化工业 包括海盐化工、海水化工、海藻化工及海洋石油化工的化工产品生产活动。

13. 海洋生物医药业 是指以海洋生物为原料或提取有效成分，进行海洋药品与海洋保健品的生产加工及制造活动。

14. 海洋工程建筑业 是指在海上、海底和海岸所进行的用于海洋生产、交通、娱乐、防护等用途的建筑工程施工及其准备活动；包括海港建筑、滨海电站建筑、海岸堤坝建筑、海洋隧道桥梁建筑、海上油气田陆地终端及处理设施建造、海底线路管道和设备安装，不包括各部门、各地区的房屋建筑及房屋装修工程。

15. 海洋电力业 是指在沿海地区利用海洋能、海洋风能进行的电力生产活动。不包括沿海地区的火力发电和核力发电。

16. 海水利用业 是指对海水的直接利用和海水淡化活动，包括利用海水进行淡水生产和将海水应用于工业冷却用水和城市生活用水、消防用水等活动，不包括海水化学资源综合利用活动。

17. 海洋交通运输业 是指以船舶为主要工具从事海洋运输以及为海洋运输提供服务的活动，包括远洋旅客运输、沿海旅客运输、远洋货物运输、沿海货物运输、水上运输辅助活动、管道运输业、装卸搬运及其他运输服务活动。

18. 滨海旅游业 是指以海岸带、海岛及海洋各种自然景观、人文景观为依托的旅游经营、服务活动，主要包括：海洋观光游览、休闲娱乐、度假住宿、体育运动等活动。

Explanatory Notes on Main Statistical Indicators

1. Marine Economy is the summation of various types of industrial activities for developing, utilizing and protecting the ocean as well as the activities associated with there.

2. Gross Ocean Product is the short form of the gross output value of ocean economy, referring to the final result of marine economic activities of the permanent units in the coastal region within a given period calculated at the market price, and the sum total of the added values of the marine and marine-related industries.

3. Marine industry refers to the production and service activities for developing, utilizing and protecting the ocean, including major marine industries such as marine fishery industry, offshore oil and gas industry, marine mining industry, sea salt industry, marine chemical industry, marine biomedicine industry, marine electric power industry, seawater utilization industry, marine shipbuilding industry, marine engineering construction industry, marine communications and transport industry, coastal tourism etc. as well as marine scientific research, education, management and service.

4. Marine Scientific Research, Education, Management and Service refer to the activities of scientific research, education, management and service carried out in the process of developing, utilizing and protecting the ocean, including marine information service industry, marine environment monitoring and forecasting service, marine insurance and social security industry, marine scientific research, marine technological service industry, ocean geological prospecting industry, marine environmental protection industry, marine education, marine management, marine social groups and international organizations etc.

5. Marine-Related Industry refers to the lower and upper reaches enterprises that form a technical and economic link with the major marine industries, with various inputs and outputs as ties, involving marine agriculture and forestry, marine equipment manufacturing, marine-related products and materials manufacturing industry, marine-related construction and installation industry, marine wholesale and retail industry, marine-related service industry etc.

6. Marine Three Industries Chinese marine three industries are divided as follows:

Marine primary industry: refers to the marine aquatic products and marine fishery service industry in the marine fishery as well as the sectors belonging to the primary industry category in the marine-related industries.

Marine secondary industry: refers to the marine aquatic products processing industry in the marine fishery, offshore oil as gas industry, marine mining industry, sea salt industry, marine chemical industry, marine biomedicine industry, marine electric power industry, seawater utilization industry, marine shipbuilding industry, marine engineering construction industry, as well as the sectors belonging to the category of secondary industry in the marine-related industries.

Marine tertiary Industry: refers to the industries other than the marine primary and secondary industries, including marine communications and transport industry, coastal tourism, marine scientific research, education, management and service industry as well as the sectors belonging to the category of tertiary industry in the marine-related industries.

7. Marine Fishery includes mariculture, marine fishing, marine fishery service industry and marine aquatic products processing, etc.

8. Offshore Oil and Gas Industry refers to the production activities of exploring, exploiting, transporting and processing crude oil and natural gas in the ocean.

9. Marine Mining Industry includes the activities of extracting and dressing beach placers, beach soil and sand, and coal mining and deep-sea mining, etc.

10. Sea Salt Industry refers to the activity of producing the salt products with the sodium chloride as the main component by utilizing seawater, including salt mining and processing.

11. Shipbuilding Industry refers to the activity of building ocean vessels, offshore fixed and floating equipment with metals or non-metals as main materials as well as repairing and dismantling ocean vessels.

12. Marine Chemical Industry includes the production activities of chemical products of sea salt, seawater, sea algal and marine petroleum chemical industries.

13. Marine Biomedicine Industry refers to the production, processing and manufacturing activities of marine medicines and marine health care products by using marine organisms as raw materials or extracting useful components therefrom.

14. Marine Engineering Construction Industry refers to the architectural projects construction and its preparations in the sea, at the sea bottom and seacoast for such uses as marine production, transportation, recreation, protection, etc., including constructions of seaports, coastal power stations, coastal dykes, marine tunnels and bridges, building of land terminals of offshore oil and gas fields as well as processing facilities, and installation of submarine pipelines and equipment, but not including the projects of house building and renovation.

15. Marine Electric Power Industry refers to the activities of generating electric power in the coastal region by making use of ocean energies and ocean wind energy. It does not include the thermal and nuclear power generation in the coastal area.

16. Seawater Utilization Industry refers to the activities of direct use of sea water and seawater desalination, including those of carrying out freshwater production and applying the seawater as water for industrial cooling, urban domestic water, water for fire fighting etc., but not including the activity of the multipurpose use of seawater chemical resources.

17. Marine Communications and Transport Industry refers to the activities of carrying out and serving the sea transportation with vessels as main vehicles, including ocean-going passengers transportation, coastal passengers transportation, ocean-going cargo transportation, coastal cargo transportation, auxiliary activities of water transportation, pipeline transportation, loading, unloading and transport as well as other transportation service activities.

18. Coastal Tourism refers to the tourist business and service activities with the backing of coastal zone, sea islands as well as a variety of natural and human landscapes of the ocean, mainly including marine sightseeing, living a life of leisure and recreation, going on vocation and getting accommodation, sports, etc.

3 主要海洋产业活动

Major Marine Industrial Activities

3-1 全国海水产品产量（按产品类别分）
Production of National Marine Products (by Product Category)

单位：万吨 (10000t)

项　目　Item	2015	2016	2017
海水养殖产量 **Mariculture Production**	**1796.56**	**1915.31**	**2000.70**
鱼类　Fish			
甲壳类　Crustacea			
贝类　Shellfish			
藻类　Algae			
其他　Others			
海洋捕捞产量 **Marine Catches**	**1216.81**	**1187.20**	**1112.42**
鱼类　Fish			
甲壳类　Crustacea			
贝类　Shellfish			
藻类　Algae			
头足类　Cephalopoda			
其他　Others			
远洋渔业产量 **Deep-sea Fishing Production**	**218.93**	**198.75**	**208.62**

3-2 全国海水产品产量（按地区分）（2017年）
Production of National Marine Products (by Regions) (2017)

单位：吨 (t)

地 区 Region	海水养殖产量 Mariculture Production	海洋捕捞产量 Marine Catches	远洋渔业产量 Deep-sea Fishing Production
全国总计 National Total	**20006973**	**11124203**	**2086200**
天 津 Tianjin	9172	27517	11900
河 北 Hebei	529158	234049	48200
辽 宁 Liaoning	3081374	552000	285400
上 海 Shanghai		14801	129900
江 苏 Jiangsu	930759	530322	26200
浙 江 Zhejiang	1162558	3093263	467900
福 建 Fujian	4453172	1743208	428200
山 东 Shandong	5190836	1749591	431300
广 东 Guangdong	3029070	1441363	47700
广 西 Guangxi	1299352	610758	8900
海 南 Hainan	321522	1127331	

3-3 沿海地区海洋原油产量
Output of Offshore Crude Oil by Coastal Regions

单位：万吨 (10000 t)

地 区 Region	2015	2016	2017
合 计 Total	**5416.35**	**5161.88**	**4886.34**
天 津 Tianjin	3113.82	2923.40	2757.68
河 北 Hebei	219.84	181.65	173.52
辽 宁 Liaoning	52.99	54.92	55.51
上 海 Shanghai	32.52	36.52	39.53
山 东 Shandong	309.20	312.40	321.23
广 东 Guangdong	1687.98	1652.99	1538.87

3-4 沿海地区海洋天然气产量
Output of Offshore Natural Gas by Coastal Regions

单位：万立方米 (10000 m^3)

地 区 Region	2015	2016	2017
合 计 Total	**1472400**	**1288604**	**1395462**
天 津 Tianjin	289771	271851	284305
河 北 Hebei	77634	55899	46341
辽 宁 Liaoning	1991	2278	1818
上 海 Shanghai	123977	151140	154570
山 东 Shandong	11676	10560	11126
广 东 Guangdong	967351	796876	897302

3-5 海洋原油出口量及创汇额
Export Volume of and Foreign-Exchange Earnings from Offshore Crude Oil by Coastal Regions

单位：万吨，万美元 (10000 t, 10000 USD)

地 区 Region	2015		2016		2017	
	出口量 Export Volume	创汇额 Foreign-Exchange Earnings	出口量 Export Volume	创汇额 Foreign-Exchange Earnings	出口量 Export Volume	创汇额 Foreign-Exchange Earnings
合 计 Total	**9**	**2946**				
天 津 Tianjin	9	2946				

3-6 海洋原油产量、出口量占全国原油产量、出口量比重
Proportion of Offshore Crude Oil Production and Export Volume in the National Total

单位：% (%)

年 份 Year	海洋原油产量占全国原油产量比重 Proportion of Offshore Crude Oil Production in the National Total	海洋原油出口量占全国原油出口量比重 Proportion of Offshore Crude Oil Export Volume in the National Total
2001	13.07	46.26
2002	14.40	54.95
2003	15.01	60.77
2004	16.16	83.37
2005	17.51	84.18
2006	17.54	89.53
2007	17.06	71.53
2008	17.96	76.46
2009	19.52	26.61
2010	23.27	24.38
2011	21.94	15.87
2012	21.42	12.43
2013	21.68	10.33
2014	21.82	0.00
2015	25.24	3.07
2016	25.85	
2017	25.52	

3-7 沿海地区海洋矿业产量
Output of Marine Mining Industry by Coastal Regions

单位：万吨 (10000 t)

地 区 Region	2015	2016	2017
合 计 Total	**4821.3**	**5167.2**	**2557.8**
浙 江 Zhejiang	2591.1	2324.7	
福 建 Fujian	204.0	1299.4	1398.5
山 东 Shandong	1327.7	1209.6	1055.3
广 西 Guangxi	494.4	333.5	100.0
海 南 Hainan	204.1	0.0	0.0

注：数据为沿海地区部分海洋矿业生产汇总数据。

Note: The data are collected from the production of part of the marine mining industry in the coastal region.

3-8 沿海地区海盐产量
Output of Sea Salt by Coastal Regions

单位：万吨 (10000 t)

地　区 Region	2015	2016	2017
合　计 Total	**3139.0**	**2839.7**	**3563.8**
天　津 Tianjin	168.7	202.7	182.3
河　北 Hebei	344.7	351.6	329.6
辽　宁 Liaoning	139.5	102.9	
江　苏 Jiangsu	84.1	72.1	68.0
浙　江 Zhejiang	7.6	6.4	5.5
福　建 Fujian	26.5	26.3	26.2
山　东 Shandong	2345.7	2065.5	2940.0
广　东 Guangdong	12.3	4.4	5.2
广　西 Guangxi	0.8	0.1	0.0
海　南 Hainan	9.1	7.7	7.0

注：2015年为中国盐业总公司数据；2016年和2017年为沿海地区汇总数据。

Note: The data for 2015 are from the China National Salt Industry Corporation, and the data for 2016 and 2017 are collected from the coastal regions.

3-9 沿海地区海洋化工产品产量
Output of Marine Chemical Products by Coastal Regions

单位：吨　　(t)

地 区 Region	2015	2016	2017
合 计 **Total**	**17949463**	**8387237**	**14518324**
天 津 Tianjin	2910772		2395434
河 北 Hebei	22864*		
辽 宁 Liaoning	1151645	1143038	1595524
江 苏 Jiangsu	858614	383785	582901
浙 江 Zhejiang	1114938	957571	1025321
福 建 Fujian	2333389	409581	2353168
山 东 Shandong	7406541	5493262	4883815
广 东 Guangdong	2150700		
海 南 Hainan			1682161

注：数据为沿海地区部分海洋化工企业产品汇总数据；*为中国盐业总公司数据。

Note: The data are collected from the products of part of the chemical enterprises in the coastal regions.
*The data are from the China National Salt Industry Corporation.

3-10 沿海地区海洋生物医药产品产量（2017年）
Production of the Marine Biomedicine Industry by Coastal Regions (2017)

产品名称 Name	计量单位 Unit	产品产量 Output
润科ARA粉剂、油剂 Runke ARA Powder, Oil	吨 t	139000
润科DHA粉剂、油剂 Runke DHA Powder, Oil	吨 t	197000
DHA DHA	吨 t	50
维生素B1 Vitamin B1	吨 t	1965
维生素B6 Vitamin B6	吨 t	685
维生素H Vitamin H	吨 t	88
伊可新 Vit AD	吨 t	280
维生素AD滴剂胶囊 Vit AD Drips, Capsules	万粒 10000 pellets	20774
辅酶Q10 Co Q10	吨 t	505
辅酶Q10胶囊 Co Q10 Capsules	万粒 10000 pellets	2042
白血病融合基因检测试剂盒 Leukemia Fusion Gene Test Kit	盒 boxes	811
地中海贫血检测试剂盒 Thalassemin Test Kit	盒 boxes	2237
结核耐药检测试剂盒 Tuberculosis Drug Resistence Test Kit	盒 boxes	4171
乙肝胶囊 Hepatitis B Capsules	吨 t	72
海藻酸钠 Alginate	吨 t	2036
螺旋藻胶囊/片 Spirulina Capsule/Pill	万粒 10000 pellets	342
螺旋藻系列产品 Spirulina Series Products	吨 t	2805
小球藻系列产品 Chlorella Series Prodcuts	吨 t	5017
岩藻黄质CWS微囊粉 Fucoxanthin Cold Water Soluble Microcapsule Powder	吨 t	3
卡拉胶 Carrageenan	吨 t	165000
琼脂 Agar	吨 t	232661
氨糖美辛片 Glucoscmine Indomethasin Tablets	万片 10000 pills	747
高纯硫酸氨基葡萄糖胶囊 High Purity Glucosamine Sulfate Capsule	万瓶 10000 bottles	36
氨糖软骨素胶囊 Glucosamine Chondroitin Capsule	盒 boxes	50948
氨基葡萄糖盐酸盐 Glucosamine Hydrochloride	吨 t	148
氨基葡萄糖系列 Glucosamine Series	吨 t	2543
几丁聚糖胶囊 Chitosan Capsule	吨 t	3
甲壳素 Chitin	吨 t	601
壳聚糖胶囊 Chitosan Capsule	瓶 bottles	61902
速溶雨生红球藻虾青素粉（固体饮料） Haematococcus Pluvialis and Astaxanthin Instant Powder (Solid Drinks)	盒 boxes	22941
虾青素微粒 Astaxanthin Beadlets	千克 kg	84

3-10 续表 continued

产品名称 Name	计量单位 Unit	产品产量 Output
虾青素CWS微囊粉 Astaxanthin CWS Microcapsule Powder	吨 t	1
虾青素油树脂 Astaxanthin Oleoresin	吨 t	1
鱼油 Fish Oil	吨 t	3978
浓缩鱼油 Concentrated Fish Oil	吨 t	2072
鱼油微囊 Fish Oil Microcapsule	吨 t	3700
鲑鱼油 Salmon Oil	粒 pellets	1900
鲛鱼油胶丸 Shark Oil Pellet	万粒 10000 pellets	40
鱼油软胶囊 Fish Oil Softgel	万瓶 10000 bottles	39
鱼油软胶囊 Fish Oil Softgel	万粒 10000 pellets	236
深海鱼油软胶囊 Deep Sea Fish Oil Capsule	万粒 10000 pellets	204
深海鱼油软胶囊300 Deep Sea Fish Oil Capsule 300	粒 pellets	25709600
天然深海鱼油软胶囊500 Natural Deep Sea Fish Oil Capsule 500	粒 pellets	243300
鱼油磷脂红花籽油软胶囊 Fish Oil Phospholipid Safflower Sead Oil Softgel	粒 pellets	720
鱼蛋白蔷薇片 Fish Protein Rose Slice	粒 pellets	63360
海洋鱼蛋白片 Marine Fish Protein Slice	粒 pellets	229200
深海鱼胶原蛋白粉 Deep Sea Fish Collagen Powder	万罐 10000 cans	2
胶原蛋白 Collagen	吨 t	298
脑元神软胶囊 Naoyunshen Softgel	万粒 10000 pellets	604
角鲨烯胶囊 Squalene Capsule	万粒 10000 pellets	15
角鲨烯软胶囊 Squalene Softgel	万粒 10000 pellets	14
鲨肝灵软胶囊 Shaganling Softgel	万粒 10000 pellets	35
鲨鱼肝油胶丸 Shark Liver Oil Capsule	万粒 10000 pellets	125
鲨鱼软骨胶囊 Shark Cartilage Capsule	万粒 10000 pellets	70
鲨鱼软骨提取物60% Shark Cartilage Extrait 60%	吨 t	1
鲨鱼软骨罐头 Shark Cartilage Can	吨 t	300000
硫酸软骨素 Chondroitin Sulfate	吨 t	954
骨制品（骨油、骨粉） Bone Products (Bone Oil, Bone Heal)	吨 t	237
海参口服液 Sea Cucumber Oral Liquid	吨 t	64830
海参肽胶囊 Sea Cucumber Peptide Capsule	吨 t	6
伤科接骨片 Traumatology Bone Graft	吨 t	251
肤疹宁软膏 Skin Rash Ointment	万支 10000 bottles	5
消毒剂 Disinfectant	吨 t	108
珍珠粉（外用） Pearl Powder (for external Use)	万袋 10000 bags	66
珍珠粉 Pearl Powder	盒 boxes	26687
珍珠粉中药饮片 Pearl Powder Decoction Pieces	盒 boxes	41259
珍珠护肤品 Pearl Skin Care Products	万瓶 10000 bottles	799

注：数据为沿海地区部分海洋生物医药产品汇总数据。

Note: The data are collected from the products of part of the marine biomedicine enterprises in the coastal regions.

3-11 沿海地区海洋修造船完工量（2017年）
Completed Quantity of Ships Repaired and Built by Coastal Regions (2017)

部门和地区 Sector and Region	修船完工量 （艘） Ships Repaired (unit)	造船完工量 Ships Built	
		艘 (unit)	万载重吨/综合吨 (10000 DWT/CT)
合　计　Total	**12446**	**2004**	**3555.2**
其　中: Including:			
中国船舶工业集团有限公司 CSSC	743	146	1294.2
中国船舶重工集团有限公司 CSIC	627	82	614.7
按地区分: by Regions:			
天　津 Tianjin	148	5	54.4
河　北 Hebei	49	81	0.4
辽　宁* Liaoning	308	47	376.7
上　海 Shanghai	973	80	989.8
江　苏 Jiangsu	269	240	1412.4
浙　江 Zhejiang	6118	591	61.6
福　建 Fujian	2536	519	55.7
山　东 Shandong	1656	188	248.0
广　东 Guangdong	341	65	354.0
广　西 Guangxi	48	19	0.6
海　南 Hainan		169	1.6

注：*为中国船舶重工集团有限公司数据。

Note: * The data are from the CSIC.

3-12 沿海地区海洋货物运输量和周转量（2017年）
Volume of Maritime Goods Transported and Turnover by Coastal Regions (2017)

单位：万吨，亿吨·公里 (10000 t, 100 million t-km)

地 区 Region	货运量 Volume of Goods Transported	沿 海 Coastal	远 洋 Oceangoing	货物周转量 Volume of Goods Turnover	沿 海 Coastal	远 洋 Oceangoing
合 计 Total	**297318**	**221288**	**76030**	**83663**	**28579**	**55084**
天 津 Tianjin	8345	8338	6	1291	1290	1
河 北 Hebei	4413	3379	1034	1204	568	636
辽 宁 Liaoning	14122	6589	7533	8609	903	7706
上 海 Shanghai	54708	30837	23871	24657	4380	20276
江 苏 Jiangsu	24154	19038	5116	4356	2071	2285
浙 江 Zhejiang	64863	63352	1511	7736	7203	532
福 建 Fujian	30344	27972	2401	5413	4618	795
山 东 Shandong	12375	10875	1500	1586	788	798
广 东 Guangdong	55010	25486	29524	23380	4120	19259
广 西 Guangxi	6451	5835	616	792	766	26
海 南 Hainan	9165	8942	223	771	710	61
其 他 Others	13369	10674	2695	3871	1161	2710

3-13 沿海地区海洋旅客运输量和周转量（2017年）
Volume of Maritime Passenger Traffic and Turnover by Coastal Regions (2017)

单位：万人，亿人·公里 (10000 persons, 100 million person-km)

地　区 Region	客运量 Passenger Traffic	沿 海 Coastal	远 洋 Oceangoing	旅客周转量 Passenger Turnover Volume	沿 海 Coastal	远 洋 Oceangoing
合　计 Total	**11764**	**10658**	**1106**	**43.4**	**30.9**	**12.5**
天　津 Tianjin	0	0	0	0.0	0.0	0.0
河　北 Hebei	2	0	2	0.1	0.0	0.1
辽　宁 Liaoning	552	543	9	6.1	5.6	0.4
上　海 Shanghai	448	448	0	0.8	0.7	0.1
江　苏 Jiangsu	14	0	14	1.1	0.0	1.1
浙　江 Zhejiang	3169	3169	0	5.3	5.3	0.0
福　建 Fujian	1670	1545	125	2.4	1.8	0.6
山　东 Shandong	1479	1389	90	11.8	7.7	4.1
广　东 Guangdong	2303	1439	863	10.0	4.1	5.9
广　西 Guangxi	381	379	2	2.0	1.9	0.1
海　南 Hainan	1746	1746	0	3.8	3.8	0.0

3-14 沿海港口客货吞吐量（2017年）
Volume of Passenger and Freight Handled at Coastal Seaports (2017)

单位：万吨，万人　　(10000 t, 10000 persons)

地 区 Region	货物吞吐量 Cargo Handled	#外 贸 Foreign Trade	旅客吞吐量 Passenger Leaving & Throught	#离 港 Leaving
合 计 Total	**905698**	**365487**	**8669**	**4314**
天 津 Tianjin	50056	28045	97	49
河 北 Hebei	108868	34593	2	1
辽 宁 Liaoning	112558	26546	587	300
上 海 Shanghai	70542	41043	355	177
江 苏 Jiangsu	29621	14498	14	7
浙 江 Zhejiang	125744	49946	691	343
福 建 Fujian	51995	20350	890	449
山 东 Shandong	151571	79988	1446	622
广 东 Guangdong	164408	55116	3061	1598
广 西 Guangxi	21862	12004	23	12
海 南 Hainan	18473	3359	1503	756

3-15 沿海地区水路国际标准集装箱运量
Volume of International Standardized Containers Traffic by Coastal Regions

单位：万标准箱，万吨 (10000 TEU, 10000 t)

地 区 Region	2015		2016		2017	
	箱 数 Number of Containers	重 量 Weight	箱 数 Number of Containers	重 量 Weight	箱 数 Number of Containers	重 量 Weight
合 计 Total	**5171**	**64869**	**5736**	**67420**	**6257**	**77982**
天 津 Tianjin	33	626	49	466	16	186
河 北 Hebei	5	71	4	58	4	65
辽 宁 Liaoning	41	411	45	490	53	611
上 海 Shanghai	1675	23793	2343	28746	2782	34633
江 苏 Jiangsu	464	4393	499	4703	546	5237
浙 江 Zhejiang	292	3769	322	4778	403	5587
福 建 Fujian	512	7968	513	7977	555	8661
山 东 Shandong	74	792	83	878	66	903
广 东 Guangdong	1126	9399	1167	9881	996	10448
广 西 Guangxi	271	4767	309	5141	337	6029
海 南 Hainan	95	1567	64	756	104	1428
其 他 Others	583	7313	339	3546	395	4193

3-16 沿海港口国际标准集装箱吞吐量
International Standardized Containers Handled at Coastal Seaports

单位：万标准箱，万吨 (10000 TEU, 10000 t)

地 区 Region	2015		2016		2017	
	箱 数 Number of Containers	重 量 Weight	箱 数 Number of Containers	重 量 Weight	箱 数 Number of Containers	重 量 Weight
合 计 Total	**18908**	**217788**	**19590**	**228980**	**21099**	**246735**
天 津 Tianjin	1411	15492	1452	15691	1507	16444
河 北 Hebei	253	3588	305	4202	374	4984
辽 宁 Liaoning	1838	31044	1880	31725	1950	33019
上 海 Shanghai	3654	35850	3713	36736	4023	39759
江 苏 Jiangsu	518	5091	490	4991	492	4902
浙 江 Zhejiang	2257	22914	2362	24439	2687	27638
福 建 Fujian	1364	17754	1440	18655	1565	20192
山 东 Shandong	2402	26851	2509	29090	2560	29567
广 东 Guangdong	4915	53957	5094	57124	5504	62130
广 西 Guangxi	142	2549	179	3335	228	4414
海 南 Hainan	154	2698	165	2991	209	3685

3-17 沿海城市国内旅游人数
Domestic Visitors by Coastal Cities

单位：万人·次 (10000 person-times)

城 市 City		2015	2016	2017
合 计	**Total**	**208504**	**199236**	**293500**
天 津	**Tianjin**	**17059**	**12036**	**20769**
河 北	**Hebei**	**7805**	**8658**	**12944**
唐 山	Tangshan	3399	4469	5591
秦皇岛	Qinhuangdao	3344	4189	5224
沧 州	Cangzhou	1062		1829
辽 宁	**Liaoning**	**18348**	**20723**	**23097**
大 连	Dalian	6828	7634	8410
丹 东	Dandong	3528	4015	4534
锦 州	Jinzhou	2070	2347	2621
营 口	Yingkou	2105	2400	2676
盘 锦	Panjin	1993	2262	2625
葫芦岛	Huludao	1824	2065	2231
上 海	**Shanghai**	**27569**	**29621**	**32845**
江 苏	**Jiangsu**	**8336**	**9377**	**10558**
南 通	Nantong	3387	3792	4247
连云港	Lianyungang	2683	3011	3384
盐 城	Yancheng	2266	2574	2927
浙 江	**Zhejiang**	**52312**	**61318**	**71946**
杭 州	Hangzhou	12040	13696	15884
宁 波	Ningbo	7920	9198	10910
温 州	Wenzhou	7576	8824	10237
嘉 兴	Jiaxing	6310	7823	9143
绍 兴	Shaoxing	7203	8288	9541
舟 山	Zhoushan	3844	4577	5473
台 州	Taizhou	7419	8912	10758
福 建	**Fujian**	**18769**	**21826**	**28032**
福 州	Fuzhou	4823	5414	6606
厦 门	Xiamen	4267	4904	7444
莆 田	Putian	1949	2316	2796
泉 州	Quanzhou	3705	4380	5329
漳 州	Zhangzhou	2190	2606	3208
宁 德	Ningde	1835	2206	2649

注：数据来源于《中国省市经济发展年鉴》。

Note：The data come from the *China Provinces And Cities Economy Development Yearbook.*

3-17 续表 continued

城　市 City		2015	2016	2017
山　东	**Shandong**	**28879**		**34628**
青　岛	Qingdao	7322		8672
东　营	Dongying	1378		1667
烟　台	Yantai	5942		7094
潍　坊	Weifang	5578		6771
威　海	Weihai	3542		4263
日　照	Rizhao	3727		4470
滨　州	Binzhou	1390		1691
广　东	**Guangdong**	**24331**	**26911**	**47857**
广　州	Guangzhou	4854	5079	5375
深　圳	Shenzhen	4157	4525	9825
珠　海	Zhuhai	1709	1909	3481
汕　头	Shantou	1426	1605	3247
江　门	Jiangmen	1543	1778	5204
湛　江	Zhanjiang	1730	1912	4306
茂　名	Maoming	708	840	1092
惠　州	Huizhou	1644	1805	2000
汕　尾	Shanwei	724	785	839
阳　江	Yangjiang	1024	1169	1311
东　莞	Dongguan	1624	1754	3738
中　山	Zhongshan	927	1056	1267
潮　州	Chaozhou	859	1016	1463
揭　阳	Jieyang	1402	1678	4709
广　西	**Guangxi**	**2423**	**5843**	**7650**
北　海	Beihai		2473	3070
防城港	Fangchenggang	1346	1569	2016
钦　州	Qinzhou	1077	1801	2564
海　南	**Hainan**	**2673**	**2923**	**4174**
海　口	Haikou	1213	1316	2410
三　亚	Sanya	1460	1607	1762
三　沙	Sansha			2

3-18 主要沿海城市接待入境过夜游客人数
Number of Inbound Tourists Received by Major Coastal Cities

单位：人·次 (person-time)

城 市	City	2015	2016	2017
合 计	**Total**	**40640543**	**44374540**	**47585633**
天 津	Tianjin	784766	824313	792094
秦皇岛	Qinhuangdao	157140	145700	153307
大 连	Dalian	984647	1044100	1063938
上 海	Shanghai	6535887	6904270	7193302
南 通	Nantong	172999	180156	185745
连云港	Lianyungang	20345	22624	26140
杭 州	Hangzhou	1417404	1580900	2040369
宁 波	Ningbo	723960	829045	890726
温 州	Wenzhou	482081	531212	585735
福 州	Fuzhou	551572	1066910	1290575
厦 门	Xiamen	1273178	2313099	2517280
泉 州	Quanzhou	663824	1250778	1383202
漳 州	Zhangzhou	292426	554700	610788
青 岛	Qingdao	880080	928297	1216971
烟 台	Yantai	382398	408522	619339
威 海	Weihai	314224	330243	414427
广 州	Guangzhou	8035800	8618800	9004800
深 圳	Shenzhen	12187100	11711700	12070100
珠 海	Zhuhai	3095100	3172300	3182500
汕 头	Shantou	211500	243900	291000
湛 江	Zhanjiang	266600	370500	372100
中 山	Zhongshan	598300	621600	661100
北 海	Beihai	129053	135536	145410
海 口	Haikou	121975	136478	181887
三 亚	Sanya	358184	448857	692798

3-19 主要沿海城市接待入境过夜游客情况（2017年）
Breakdown of Inbound Tourists Received by Major Coastal Cities (2017)

单位：人·次，人·天 (person-time, night)

城市 City		外国人 Foreigners		香港同胞 Hong Kong Compatriots	
		人次数 Arrivals	人天数 Nights	人次数 Arrivals	人天数 Nights
天　津	Tianjin	685332	12530773	53573	1331935
秦皇岛	Qinhuangdao	140976	907797	5805	28273
大　连	Dalian	903726	1877957	76290	153594
上　海	Shanghai	5894849	19453001	516104	1703143
南　通	Nantong	152124	432660	7756	17213
连云港	Lianyungang	21541	69756	675	1406
杭　州	Hangzhou	1489721	4016411	193840	516838
宁　波	Ningbo	703180	1548262	80137	165322
温　州	Wenzhou	364515	801069	78909	176397
福　州	Fuzhou	811405	5110145	145113	928288
厦　门	Xiamen	1255929	7036207	307623	1531980
泉　州	Quanzhou	363689	2160777	655728	3316118
漳　州	Zhangzhou	157232	866358	155631	668298
青　岛	Qingdao	907925	2906513	140718	402866
烟　台	Yantai	490216	1841864	37798	135921
威　海	Weihai	379670	1123300	4287	11117
广　州	Guangzhou	3457400	11541300	4371700	11357600
深　圳	Shenzhen	1775600	3518500	9822800	21644500
珠　海	Zhuhai	510700	1060500	1155600	1910100
汕　头	Shantou	169600	432600	107200	156500
湛　江	Zhanjiang	207300	435100	133500	221300
中　山	Zhongshan	127000	454300	389200	1113200
北　海	Beihai	76008	138418	44568	79257
海　口	Haikou	97703	182957	27004	42378
三　亚	Sanya	539383	2198432	78803	170309

3-19 续表 continued

城 市 City		澳门同胞 Macao Compatriots		台湾同胞 Taiwan Compatriots	
		人次数 Arrivals	人天数 Nights	人次数 Arrivals	人天数 Nights
天 津	Tianjin	3543	88597	49646	1298043
秦皇岛	Qinhuangdao	224	1339	6302	34346
大 连	Dalian	2322	5653	81600	149504
上 海	Shanghai	28424	85272	753925	2638738
南 通	Nantong	340	721	25525	80700
连云港	Lianyungang	21	43	3903	15915
杭 州	Hangzhou	21467	48935	335341	830844
宁 波	Ningbo	19128	42647	88280	174547
温 州	Wenzhou	47483	188300	94828	199515
福 州	Fuzhou	14775	84367	319282	1537494
厦 门	Xiamen	16343	79906	937385	4650228
泉 州	Quanzhou	70489	332037	293296	1346847
漳 州	Zhangzhou	19877	86519	278048	1062355
青 岛	Qingdao	37923	131587	130405	402207
烟 台	Yantai	19582	62739	71743	224664
威 海	Weihai	1100	2819	29370	79542
广 州	Guangzhou	544100	1518400	631600	1794400
深 圳	Shenzhen	59200	136000	412500	842600
珠 海	Zhuhai	887100	1950600	629100	1107700
汕 头	Shantou	2000	3200	12200	22400
湛 江	Zhanjiang	10100	21100	21300	46400
中 山	Zhongshan	85500	195500	59400	202800
北 海	Beihai	9091	17812	15743	27648
海 口	Haikou	2398	3550	54782	82370
三 亚	Sanya	7869	17763	66743	143193

主要统计指标解释

1. 海洋捕捞产量　凡是从海洋里捕捞的天然生长的水产品产量为海洋捕捞产量。

2. 海水养殖产量　凡是从人工投放苗种或天然纳苗并进行人工饲养管理的海水养殖水域中捕捞的水产品产量为海水养殖产量。

3. 远洋渔业产量　由各远洋渔业企业和各生产单位按我国远洋渔业项目管理办法组织的远洋渔船（队）在非我国管辖海域（外国专属经济区水域或公海）捕捞的水产品产量。中外合资、合作渔船捕捞的水产品只统计按协议应属于中方所有的部分。

4. 原油产量　是按净原油量来计算的，能直接用于销售和生产自用的原油量。目前海洋石油系统原油产量计算方法采用倒算法。

原油产量=销售量+期末库存量-期初库存量+海上平台及陆地终端处理厂自用量

5. 天然气产量　指进入集输管网的销售量和就地利用的全部气量。

天然气产量=外输（销）量+企业自用量

6. 造船综合吨　等于以计量单位载重吨和满载排水量吨的民用船舶的吨位数之和。

7. 货运量　指经船舶实际运送的货物重量。

8. 货物周转量　指实际运送的货物与其运送距离的乘积。

9. 集装箱运量　既包含货重，也包含箱重。箱重系指承运租用的空箱重量凡有运费收入的空箱，其重量应统计为运量，按空箱 1 吨为货运量 1 吨计算；若无收入，所承运的空箱一律不作运量统计。

10. 旅客周转量　指实际运送的旅客人数与其运送距离的乘积。

11. 接待人次数　指报告期内我国接待游客人数。游客按出游地分为入境游客和国内游客，按出游时间分为旅游者（过夜游客）和一日游游客（不过夜游客）。

12. 接待人天数　指过夜旅游者的停留天数。

13. 外国人　指外国国籍的人，加入外籍的中国血统华人也计入外国人。

14. 港澳台同胞　指居住在我国香港特别行政区、澳门特别行政区和台湾省的中国同胞。

Explanatory Notes on Main Statistical Indicators

1. Marine Catches　refers to the output of the naturally growing aquatic products caught from the sea.

2. Mariculture Production　refers to the output of aquatic products whose young are artificially released or naturally collected, and raised and managed artificially, and which are caught from the waters of mariculture.

3. Deep-Sea Fishing Production　refers to the output of aquatic products caught in the non-Chinese jurisdictional sea areas (foreign EEE or high sea) by the distant fishing vessels (fleet) organized by various distant fishing businesses and production units according to the management measures for the China distant fishing projects. The aquatic products caught by the Chinese-foreign joint ventures' and cooperative fishing vessels are counted only for the part owned by the Chinese side according to the agreement.

4. Output of Crude Oil is calculated on the basis of the net amount of crude oil, i.e., the amount of crude oil that may be directly used for sale and for the production itself.

Output of crude oil = Volume of sales + Inventory at the end of the period - Inventory at the beginning of the period +Amount for self-use on the platforms and in the terminal processing plants on land

5. Output of Natural Gas refers to the total gas volume of the sales volume entering the oil collecting and transport pipeline network and that used locally.

Output of natural gas = Volume of sales or transport to other areas + Volume used by the enterprise itself

6. Comprehensive Tonnages of Shipbuilding refers to the sum of tonnage of civilian vessels with the deadweight capacity and full-load displacement as measured.

7. Freight Traffic refers to the weight of cargoes actually transported by vessels.

8. Cargoes Turnover Volume refers to the product of the actually transported cargoes and the transport distance.

9. Freight Volume of Containers includes the weight of both cargoes and container boxes. The weight of container boxes refers to the weight of empty containers rented for transport or having freight income and should be included in the freight volume, one ton of empty boxes equalling to one ton of freight volume. The empty boxes which have no income for transportation are not included in the freight volume.

10. Passenger Turnover Volume refers to the product of the number of passengers actually transported and the shipping distance.

11. Number of Person-Times Received refers to the number of visitors received by China in the period reported. Visitors are divided into inbound visitors and domestic visitors by origin of the travel, and tourists (overnight visitors) and one-day visitors (non overnight visitors) by their length of stay.

12. Number of the Days of Stay refers to the number of the days of stay of tourists.

13. Foreigners refers to the people with foreign nationality, including foreign nationals of Chinese descent.

14. Compatriots from Hong Kong, Macao and Taiwan Province refers to the Chinese compatriots living in the Hong Kong Special Administrative Region, the Macao Special Administrative Region and Taiwan Province.

4

主要海洋产业生产能力

Production Capacity of Major Marine Industries

4-1 沿海地区渔港情况（2017年）
Fishing Ports in the Coastal Regions (2017)

单位：个 (unit)

地 区 Region		合计 Total	中心渔港 Central Fishing Port	一级渔港 Grade 1 Fishing Port
合 计	**Total**	**134**	**58**	**70**
天 津	Tianjin			
河 北	Hebei	8	3	4
辽 宁	Liaoning	10	3	6
上 海	Shanghai	1		1
江 苏	Jiangsu	12	6	5
浙 江	Zhejiang	22	9	13
福 建	Fujian	22	8	13
山 东	Shandong	21	11	9
广 东	Guangdong	19	8	11
广 西	Guangxi	9	4	4
海 南	Hainan	10	6	4

注：数据来源于《2018中国渔业统计年鉴》（省级数据中不包含计划单列市数据）。

Note: The data come from the *China Fishery Statistical Yearbook 2018* (The data of the provinces don't include the data of the cities specifically designated in the state plan).

4-2 沿海地区海水养殖面积
Mariculture Area by Coastal Regions

单位：公顷 (hm²)

地 区 Region	2015	2016	2017
合 计 **Total**	**2317763**	**2098103**	**2084076**
天 津 Tianjin	3165	8999	3206
河 北 Hebei	117533	124800	107583
辽 宁 Liaoning	933068	698400	698400
上 海 Shanghai			
江 苏 Jiangsu	181829	185480	192390
浙 江 Zhejiang	85881	78701	75954
福 建 Fujian	166075	153000	155739
山 东 Shandong	563198	604800	610377
广 东 Guangdong	194861	166200	161690
广 西 Guangxi	55015	45400	47022
海 南 Hainan	17138	32323	31715

4-3 海洋油气勘探情况（2017年）
Exploration of Offshore Oil and Gas (2017)

地　区 Region	地震测线 Seismic Line		钻井（口） Drilling (well)	
	二维 （千米） Two Dimensions (km)	三维 （平方千米） Three Dimensions (km^2)	预探井 Wildcat Wells	评价井 Appraisal Wells
合　计 **Total**	**4417**	**11363**	**81**	**83**
天　津 Tianjin		742	31	45
河　北 Hebei			14	
上　海 Shanghai		1934	2	1
山　东 Shandong		300	1	6
广　东 Guangdong	4417	8387	33	19
其中：合作 Including: Cooperative	2169	1711	1	

4-4 海洋油气生产井情况（2017年）
Survey of Offshore Oil and Gas Production Wells (2017)

单位：口　　　　(well)

地　区 Region	合　计 Total	采油井 Oil Wells	采气井 Gas Wells	注水井 Injection Wells	其他井 Others
合　计　**Total**	**7638**	**5523**	**337**	**1778**	
天　津　Tianjin	4110	2947	117	1046	
河　北　Hebei	1402	1012	7	383	
辽　宁　Liaoning	298	246	6	46	
上　海　Shanghai	93	21	72		
山　东　Shandong	800	528	7	265	
广　东　Guangdong	935	769	128	38	

4-5 沿海地区盐田面积和海盐生产能力
Salt Pan Area and Sea Salt Production Capacity by Coastal Regions

地　区 Region	盐田总面积（公顷） Total Area of Salt Pan (hm^2)		生产面积（公顷） Production Area (hm^2)		年末海盐生产能力（万吨） Year-End Capacity of Sea Salt Production (10000 t)	
	2016	2017	2016	2017	2016	2017
合　计 Total	**369566**	**294284**	**303258**	**250239**	**3750**	**2011**
天 津 Tianjin	26895	26335	26200	23668	161	161
河 北 Hebei	69840	64840	59650	54910	414	342
辽 宁 Liaoning	30744		26085		134	
江 苏 Jiangsu	39175	2633	17213	1860	70	21
浙 江 Zhejiang	1683	846	1499	769	8	3
福 建 Fujian	3998	3998	3596	3596	26	28
山 东 Shandong	193017	190500	165365	161250	2907	1434
广 东 Guangdong		2130		1538	12	6
广 西 Guangxi	711		373			
海 南 Hainan	3503	3002	3277	2648	18	17

4-6 海上风电项目情况（2017年）
Projects of Offshore Wind Power (2017)

地 区 Region	海上风电新增项目情况 New Offshore Wind Power Projects		海上风电已安装项目情况 Installed Offshore Wind Power Projects	
	新增装机数量（台） New Installed Number (unit)	新增装机容量（兆瓦） New Installed Capacity (MW)	累计装机数量（台） Cumulative Capacity (unit)	累计装机容量（兆瓦） Cumulative Installed Capacity (MW)
天 津 Tianjin	0	0	18	27.0
河 北 Hebei	9	36	9	36.0
辽 宁 Liaoning	2	6	3	7.5
上 海 Shanghai	0	0	90	305.0
江 苏 Jiangsu	267	968	625	2173.7
浙 江 Zhejiang	11	44	11	44.0
福 建 Fujian	15	65	30	136.0
山 东 Shandong	0	0	4	15.0
广 东 Guangdong	15	45	16	46.5

4-7 主要潮汐电站分布情况
Distribution of Major Tidal Power Stations

电站名称 Name	运行情况 Status of Operation	装机容量（千瓦） Installed Capacity (kW)
江厦潮汐试验电站 Jiangxia Experimental Tidal Power Station	1972年开始建造，1980年投入使用，运行至今 It began construction in 1972, was put into use in 1980, and has been in operation so far.	4100
海山潮汐电站 Haishan Tidal Power Station	1972年开始建造，1975年投入使用，运行至今 It began construction in 1972, was put into use in 1975, and has been in operation so far.	250
岳浦潮汐电站 Yuepu Tidal Power Station	1970年开始建造，1978年停止运行 It began construction in 1970, and stopped power generation in 1978.	300
白沙口潮汐电站 Baishakou Tidal Power Station	1970年开始建造，1978年投入使用，2010年停止运行 It began construction in 1970, was put into use in 1978, and stopped power generation in 2010.	960

4-8 主要海上活动船舶（2017年）
Major Vessels Operating on the Sea (2017)

类 别 Type	艘数 （艘） Number of Vessels (unit)	总吨 （万吨位） Gross Tonnage (10000 t)	净载重量 （万吨） Net Weight Tonnage (10000 t)	载客量 （客位） Passenger Capacity (seat)	总功率 （千瓦） Total Power (kW)
一 、海洋生产用船 Vessels for Marine Production					
海洋渔业船舶 Marine Fishing Vessels	232048	794 .5			14803573
远洋渔船 Ocean-going Fishing Vessels	2491				2551789
海洋油气船舶 Offshore Oil and Gas Vessels	234	311 .2	312 .4	13020	1855349
钻井平台 Drilling Vessels	49	0 .0	0 .0	5020	536704
物探船 Physical Exploration Vessels	14	7 .2	0 .0	480	70674
其 他 Others	171	304 .0	312 .4	7520	1247971
海洋运输船舶 Marine Transport Vessels	12624	8766 .2	12501 .9	244488	34026272
二、海洋科研用船 Vessels for Marine Scientific Research					
海洋地质勘探船 Marine Geological Survey Vessels	9	2 .7	1 .2	528	48942
海洋调查船 Marine Research Vessels					
中国科学院 Chinese Academy of Sciences	9	2 .0	0 .5	343	29706
自然资源部* Ministry of Natural Resources of the PRC	16	5 .6		668	78308

注：*为2016年数据。

Note: * Data for 2016.

4-9 沿海主要港口生产用码头泊位（2017年）
Berths for Productive Use at the Main Coastal Seaports (2017)

单位：米, 个 (m, unit)

港口	Seaport	码头长度 Wharf Length	泊位个数 Number of Berths	#万吨级 10000 Tons Class
合　计	**Total**	**800000**	**5324**	**1892**
丹　东	Dandong	7626	42	25
大　连	Dalian	41101	223	104
营　口	Yingkou	18975	86	61
锦　州	Jinzhou	6119	23	21
秦皇岛	Qinhuangdao	15928	72	44
黄　骅	Huanghua	9586	39	33
唐　山	Tangshan	30036	111	108
天　津	Tianjin	35478	145	117
烟　台	Yantai	32550	195	89
威　海	Weihai	15188	94	33
青　岛	Qingdao	28818	121	84
日　照	Rizhao	18479	73	63
上　海	Shanghai	72473	563	172
连云港	Lianyungang	15570	65	57
盐　城	Yancheng	8924	87	15
嘉　兴	Jiaxing	11148	85	35
宁波舟山	Ningbo Zhoushan	89667	613	171
台　州	Taizhou	13024	180	9
温　州	Wenzhou	16872	206	20

4-9 续表 continued

港口 Seaport		码头长度 Wharf Length	泊位个数 Number of Berths	#万吨级 10000 Tons Class
福州	Fuzhou	26572	187	58
莆田	Putian	6121	46	12
泉州	Quanzhou	15971	104	25
厦门	Xiamen	29658	165	76
汕头	Shantou	9627	87	19
汕尾	Shanwei	1365	12	2
惠州	Huizhou	9748	43	21
深圳	Shenzhen	31075	141	73
虎门	Humen	16430	120	30
广州	Guangzhou	49686	485	73
中山	Zhongshan	5209	67	0
珠海	Zhuhai	19194	156	28
江门	Jiangmen	12905	158	6
阳江	Yangjiang	2232	10	9
茂名	Maoming	2384	16	9
湛江	Zhanjiang	16353	119	36
广西北部湾港	Beibuwan	37359	263	86
海口	Haikou	9385	68	33
洋浦	Yangpu	8676	42	26
八所	Basuo	2488	12	9

4-10 沿海地区星级饭店基本情况（2017年）
Star Grade Hotels and Occupancies by Coastal Regions (2017)

地　区 Region		饭店数（座） Number of Hotels (unit)	客房数（间） Number of Rooms (unit)	床位数（张） Number of Beds (unit)	客房出租率（%） Room Occupancy (%)
合　计	**Total**	**4116**	**679623**	**1112485**	
天　津	Tianjin	80	16076	24437	55.06
河　北	Hebei	338	51264	91614	45.08
辽　宁	Liaoning	336	51832	85891	48.70
上　海	Shanghai	223	58658	87266	68.98
江　苏	Jiangsu	514	78845	122699	59.79
浙　江	Zhejiang	585	98358	157746	57.73
福　建	Fujian	306	52354	81902	58.44
山　东	Shandong	586	82112	146757	54.94
广　东	Guangdong	658	116114	184126	58.13
广　西	Guangxi	370	48446	86600	54.02
海　南	Hainan	120	25564	43447	63.19

4-11 沿海地区旅行社数
Number of Travel Agencies by Coastal Regions

单位：家 (unit)

地　区 Region	2015	2016	2017
合　计　Total	**14184**	**14457**	**15502**
天　津　Tianjin	400	396	475
河　北　Hebei	1360	1373	1382
辽　宁　Liaoning	1253	1258	1246
上　海　Shanghai	1225	1261	1382
江　苏　Jiangsu	2160	2241	2364
浙　江　Zhejiang	2028	2051	2216
福　建　Fujian	846	844	897
山　东　Shandong	2109	2115	2220
广　东　Guangdong	1901	2028	2450
广　西　Guangxi	539	586	598
海　南　Hainan	363	304	272

主要统计指标解释

1. 海水养殖面积 是指利用天然海水用于养殖水产品的水面面积，包括海上养殖、滩涂养殖、其他养殖。在报告期内无论是否全部收获或尚未收获其产品，均应统计在海水养殖面积中。但有些滩涂、水面不投放苗种或投放少量苗种，只进行一般管理的，不统计为养殖面积。

2. 盐田总面积 指盐田占有的全部面积。包括储卤、蒸发、保卤、结晶面积、滩内的沟、壕、池、埝、滩坨等面积及滩外的沟、壕、公路及杂地面积。

3. 盐田生产面积 指直接提供给海盐生产的面积，包括结晶面积、蒸发面积、保卤面积，滩内的沟、壕、池、埝面积及滩坨面积。

4. 年末海盐生产能力 指年末企业生产原盐的全部设备的综合平衡能力。海盐生产露天作业，受天气影响，因而计算生产能力时，成熟滩田按 10 年实际平均单位生产面积产量乘以本年成熟滩田生产面积而得，新滩田按设计能力及滩田成熟程度可能达到的产量计算。

5. 海洋渔业船舶 是指配置机器作为动力的从事海洋渔业生产和辅助渔业生产的船舶。

6. 远洋渔船 按我国远洋渔业项目管理办法在非我国管辖海域（外国专属经济区水域或公海）进行常年或季节性生产的渔船。

7. 泊位个数 是指设有系靠船舶设施，在同一时间内可供靠泊最大吨级船舶的艘数。即可靠泊一艘船舶，则计为一个泊位，余类推。泊位分码头泊位和浮筒泊位。

8. 客房数 指饭店实际可用于接待旅游者的房间数。

9. 床位数 指饭店实际可用于接待旅游者的床位数。

Explanatory Notes on Main Statistical Indicators

1. Mariculture Area refers to the area of the water surface where aquatic products are cultivated by using natural seawater, including maritime culture, tidal flat culture and other cultures. Whether or not all the products in the area have been harvested or the products have not been harvested yet in the period covered by the report, the area is included in the Mariculture Area. But some tidal flats and water surfaces where none or a small amount of the young have been released and only general management is carried out are not included in the Mariculture Area.

2. Total Area of Salt Pans refers to the total area covered by salt pans, including the area for brine storage, evaporation, brine preservation, and crystallization, the area of ditches, moats, ponds and banks within the beach as well as beach mounds, and ditches, moats, highway beyond the beach as well as the area of miscellaneous lands.

3. Area of Salt Pan Production refers to the area directly provided for sea-salt productions, including the area for crystallization, evaporation and brine preservation, the area of ditches, moats, ponds and banks within the beach as well as the area of beach mounds.

4. Year-End Capacity of Crude Salt Production refers to the integrated and balanced capacity of all equipment of the enterprise used for crude salt production at the end of the year. As sea salt production is an open-air operation, which is subject to the effect of weather, the production capacity of a matured salt pan is calculated at the productions of the actual average unit production area in ten years times the production area of the matured salt pan in the current year. The production capacity of new salt pans is calculated at the production that may be reached in the light of the designed capacity and the level of maturity of the salt pan.

5. Marine Fishing Vessels refer to the vessels equipped with machines as motive power and going for marine fishery production and auxiliary fishery production.

6. Deep Sea Fishing Vessels refer to the fishing vessels which carry out production all the year round or seasonally in the non-Chinese jurisdictional sea areas (foreign EEZ or high sea) according to the China Deep-Sea Fishing Projects Management Measures.

7. Number of Berths refers to the spaces equipped with facilities for docking ships and the number of ships of the maximum tonnage that may dock or anchor in them. A space for a ship to dock is counted as one berth and the rest are reasoned out by analogy. Berths are divided into wharf berths and buoy berths.

8. Number of Rooms refers to the number of guest rooms actually used by the hotels receiving tourists.

9. Number of Beds refers to the number of beds actually used by the hotels receiving tourists.

5

海洋科学技术

Marine Science and Technology

5-1 分行业海洋科研机构及人员情况（2017年）
Marine Scientific Research Institutions and by Personnel Industry (2017)

行 业 Industry	机构数（个） Number of Institutions (unit)	从业人员（人） Employees (person)
合 计 **Total**	**159**	**29089**
海洋基础科学研究 **Marine Basic Scientific Research**	**100**	**19057**
海洋自然科学 Marine Natural Science	60	14788
海洋社会科学 Marine Social Science	4	787
海洋农业科学 Marine Agricultural Science	35	3415
海洋生物医药 Marine Biomedicine	1	67
海洋工程技术研究 **Marine Engineering Technology Research**	**48**	**6904**
海洋化学工程技术 Marine Chemical Engineering Technology	2	178
海洋生物工程技术 Marine Bioengineering Technology	2	220

注：机构为县级以上科研机构；行业分类参照《海洋及相关产业分类》(GB/T 20794—2006)标准。
统计口径发生变化，不包括涉海企业科研机构数据（本部分其他表同）。

Note: The institutions are the scientific research institutions above the county level; The classification of industries follows the standard *Classification of Marine Industries and the Related Industries (GB/T 20794—2006).*
The statistical caliber has changed, so the data of the statistical institutions don't include those of the scientific research institutions of marine related enterprises. The same applies to the tables in Part 6.

5-1 续表 continued

行 业 Industry	机构数（个） Number of Institutions (unit)	从业人员（人） Employees (person)
海洋交通运输工程技术 Marine Communications and Transport Engineering Technology	9	1915
海洋能源开发技术 Marine Energies Development Technology	1	39
海洋环境工程技术 Marine Environmental Engineering Technology	14	1030
河口水利工程技术 Estuarine Water Conservancy Engineering Technology	17	3136
其他海洋工程技术 Other Marine Engineering Technologies	3	386
海洋信息服务业 **Marine Information Service**	**9**	**708**
海洋技术服务业 **Marine Technological Service Industry**	**2**	**2420**
海洋工程管理服务 Marine Engineering Management Service	0	0
其他海洋专业技术服务 Other Marine Professional and Technological Services	2	2420

5-2 分行业海洋科研机构科技活动人员学历构成（2017年）
Educational Background Composition of the Personnel Engaged in Scientific and Technological Activities in Marine Scientific Research Institutions by Industry (2017)

单位：人 (person)

行 业 Industry	科技活动人员 Personnel Engaged in Scientifical Activities	#博士 Doctor	#硕士 Master	#大学生 University Graduate	#大专生 College Graduate
合 计 Total	**25642**	**7811**	**8525**	**6853**	**1541**
海洋基础科学研究 Marine Basic Scientific Research	**17609**	**6527**	**5256**	**3994**	**1046**
海洋自然科学 Marine Natural Science	13830	5799	4006	2700	748
海洋社会科学 Marine Social Science	776	165	310	257	36
海洋农业科学 Marine Agricultural Science	2951	563	933	998	256
海洋生物医药 Marine Biomedicine	52	0	7	39	6
海洋工程技术研究 Marine Engineering Technology Research	**5772**	**868**	**2135**	**2322**	**385**
海洋化学工程技术 Marine Chemical Engineering Technology	88	1	23	49	13
海洋生物工程技术 Marine Bioengineering Technology	199	20	90	70	19

5-2 续表 continued

行 业 Industry	科技活动人员 Personnel Engaged in Scientifical Activities	#博士 Doctor	#硕士 Master	#大学生 University Graduate	#大专生 College Graduate
海洋交通运输工程技术 Marine Communications and Transport Engineering Technology	1665	141	772	675	62
海洋能源开发技术 Marine Energies Development Technology	37	0	11	14	12
海洋环境工程技术 Marine Environmental Engineering Technology	921	67	274	484	83
河口水利工程技术 Estuarine Water Conservancy Engineering Technology	2632	592	854	964	190
其他海洋工程技术 Other Marine Engineering Technologies	230	47	111	66	6
海洋信息服务业 Marine Information Service	**661**	**77**	**259**	**282**	**36**
海洋技术服务业 Marine Technological Service Industry	**1600**	**339**	**875**	**255**	**74**
海洋工程管理服务 Marine Engineering Management Service	0	0	0	0	0
其他海洋专业技术服务 Other Marine Professional and Technological Services	1600	339	875	255	74

5-3 分行业海洋科研机构科技活动人员职称构成（2017年）
Professional Title Composition of Personnel Engaged in Scientific and Technological Activities in the Marine Scientific Research Institutions by Industry (2017)

单位：人 (person)

行 业 Industry	科技活动人员 Personnel Engaged in Scientifical Activities	#高级职称 Senior Professional Title	#中级职称 Intermediate Professional Title	#初级职称 Junior Professional Title
合 计 Total	**25642**	**11268**	**8823**	**3225**
海洋基础科学研究 Marine Basic Scientific Research	**17609**	**7834**	**6287**	**1998**
海洋自然科学 Marine Natural Science	13830	6333	4996	1486
海洋社会科学 Marine Social Science	776	400	173	52
海洋农业科学 Marine Agricultural Science	2951	1100	1109	437
海洋生物医药 Marine Biomedicine	52	1	9	23
海洋工程技术研究 Marine Engineering Technology Research	**5772**	**2625**	**1778**	**877**
海洋化学工程技术 Marine Chemical Engineering Technology	88	20	47	18
海洋生物工程技术 Marine Bioengineering Technology	199	70	90	38

5-3 续表 continued

行 业 Industry	科技活动人员 Personnel Engaged in Scientifical Activities	#高级职称 Senior Professional Title	#中级职称 Intermediate Professional Title	#初级职称 Junior Professional Title
海洋交通运输工程技术 Marine Communications and Transport Engineering Technology	1665	608	516	310
海洋能源开发技术 Marine Energies Development Technology	37	13	19	2
海洋环境工程技术 Marine Environmental Engineering Technology	921	343	335	193
河口水利工程技术 Estuarine Water Conservancy Engineering Technology	2632	1451	698	288
其他海洋工程技术 Other Marine Engineering Technologies	230	120	73	28
海洋信息服务业 Marine Information Service	**661**	**254**	**223**	**129**
海洋技术服务业 Marine Technological Service Industry	**1600**	**555**	**535**	**221**
海洋工程管理服务 Marine Engineering Management Service	0	0	0	0
其他海洋专业技术服务 Other Marine Professional and Technological Services	1600	555	535	221

5-4 分行业海洋科研机构经费收入（2017年）

Routine Funds Receipts of the Marine Scientific Research Institutions by Industry (2017)

单位：千元 (1000 yuan)

行 业 Industry	经费收入总额 Fund Total	本年收入合计 Total Annual Income	基本建设中政府投资 Government Investment in the Capital Construction
合 计 Total	**26345457**	**24125285**	**2220172**
海洋基础科学研究 Marine Basic Scientific Research	**18558931**	**16955652**	**1603279**
海洋自然科学 Marine Natural Science	15729164	14491647	1237517
海洋社会科学 Marine Social Science	543660	489277	54383
海洋农业科学 Marine Agricultural Science	2274420	1963041	311379
海洋生物医药 Marine Biomedicine	11687	11687	0
海洋工程技术研究 Marine Engineering Technology Research	**5164760**	**4816372**	**348388**
海洋化学工程技术 Marine Chemical Engineering Technology	166376	166155	221
海洋生物工程技术 Marine Bioengineering Technology	257309	257309	0

5-4 续表 continued

行 业 Industry	经费收入总额 Fund Total	本年收入合计 Total Annual Income	基本建设中政府投资 Government Investment in the Capital Construction
海洋交通运输工程技术 Marine Communications and Transport Engineering Technology	1539121	1260254	278867
海洋能源开发技术 Marine Energies Development Technology	18982	18982	0
海洋环境工程技术 Marine Environmental Engineering Technology	698410	658410	40000
河口水利工程技术 Estuarine Water Conservancy Engineering Technology	2204270	2176649	27621
其他海洋工程技术 Other Marine Engineering Technologies	280292	278613	1679
海洋信息服务业 Marine Information Service	**663536**	**498574**	**164962**
海洋技术服务业 Marine Technological Service Industry	**1958230**	**1854687**	**103543**
海洋工程管理服务 Marine Engineering Management Service	0	0	0
其他海洋专业技术服务 Other Marine Professional and Technological Services	1958230	1854687	103543

5-5 分行业海洋科研机构科技课题情况（2017年）
Marine Science and Technology Research Projects of the Research Institutions by Industry (2017)

单位：项 (item)

行　业 Industry	课题数 Number of Research Projects	基础研究 Basic Research	应用研究 Applied Research	试验发展 Experimental Development	成果应用 Result Application	科技服务 Scientific and Technological Service
合 计 Total	**21257**	**6282**	**6781**	**3276**	**2163**	**2755**
海洋基础科学研究 Marine Basic Scientific Research	**16877**	**5617**	**5073**	**2425**	**1720**	**2042**
海洋自然科学 Marine Natural Science	13553	5331	4499	1821	920	982
海洋社会科学 Marine Social Science	1463	8	23	19	561	852
海洋农业科学 Marine Agricultural Science	1846	276	547	578	237	208
海洋生物医药 Marine Biomedicine	15	2	4	7	2	0
海洋工程技术研究 Marine Engineering Technology Research	**3303**	**276**	**1292**	**747**	**318**	**670**
海洋化学工程技术 Marine Chemical Engineering Technology	8	0	5	3	0	0
海洋生物工程技术 Marine Bioengineering Technology	152	0	15	91	37	9

5-5 续表 continued

行 业 Industry	课题数 Number of Research Projects	基础研究 Basic Research	应用研究 Applied Research	试验发展 Experimental Development	成果应用 Result Application	科技服务 Scientific and Technological Service
海洋交通运输工程技术 Marine Communications and Transport Engineering	903	121	115	159	126	382
海洋能源开发技术 Marine Energies Development	11	0	0	6	2	3
海洋环境工程技术 Marine Environmental Engineering Technology	281	17	19	76	32	137
河口水利工程技术 Estuarine Water Conservancy Engineering Technology	1862	128	1120	355	121	138
其他海洋工程技术 Other Marine Engineering Technologies	86	10	18	57	0	1
海洋信息服务业 Marine Information Service	**56**	**1**	**8**	**8**	**0**	**39**
海洋技术服务业 Marine Technological Service Industry	**1021**	**388**	**408**	**96**	**125**	**4**
海洋工程管理服务 Marine Engineering Management Service	0	0	0	0	0	0
其他海洋专业技术服务 Other Marine Professional and Technological Services	1021	388	408	96	125	4

5-6 分行业海洋科研机构科技论著情况（2017年）
Marine Scientific and Technological Theses and Works of the Research Institutions by Industry (2017)

行　业 Industry	发表科技论文（篇） Scientific Theses Published (piece)	#国外发表 Published Abroad	出版科技著作（种） Scientific and Technological Works Published (kind)
合　计 Total	**15872**	**7587**	**388**
海洋基础科学研究 Marine Basic Scientific Research	**12130**	**6049**	**261**
海洋自然科学 Marine Natural Science	9860	5609	140
海洋社会科学 Marine Social Science	646	25	84
海洋农业科学 Marine Agricultural Science	1617	415	37
海洋生物医药 Marine Biomedicine	7	0	0
海洋工程技术研究 Marine Engineering Technology Research	**2154**	**275**	**111**
海洋化学工程技术 Marine Chemical Engineering Technology	10	0	0
海洋生物工程技术 Marine Bioengineering Technology	39	1	1

5-6 续表 continued

行 业 Industry	发表科技论文（篇） Scientific Theses Published (piece)	#国外发表 Published Abroad	出版科技著作（种） Scientific and Technological Works Published (kind)
海洋交通运输工程技术 Marine Communications and Transport Engineering Technology	558	73	28
海洋能源开发技术 Marine Energies Development Technology	9	0	1
海洋环境工程技术 Marine Environmental Engineering Technology	149	19	6
河口水利工程技术 Estuarine Water Conservancy Engineering Technology	1222	182	75
其他海洋工程技术 Other Marine Engineering Technologies	167	0	0
海洋信息服务业 **Marine Information Service**	**130**	**9**	**7**
海洋技术服务业 **Marine Technological Service Industry**	**1458**	**1254**	**9**
海洋工程管理服务 Marine Engineering Management Service	0	0	0
其他海洋专业技术服务 Other Marine Professional and Technological Services	1458	1254	9

5-7 分行业科研机构科技专利情况（2017年）
Marine Scientific and Technological Patents of the Research Institutions by Industry (2017)

单位：件 (unit)

行 业 Industry	专利申请受理数 Number of Patent Applications Accepted	#发明专利 Patents for Discoveries	专利授权数 Number of Patents Granted	#发明专利 Patents for Discoveries	拥有发明专利总数 Total Number of Patents for Discoveries
合 计 Total	**4779**	**3347**	**3288**	**2242**	**10352**
海洋基础科学研究 Marine Basic Scientific Research	**2499**	**1886**	**1983**	**1379**	**7361**
海洋自然科学 Marine Natural Science	1890	1447	1495	1117	5652
海洋社会科学 Marine Social Science	23	11	30	18	34
海洋农业科学 Marine Agricultural Science	586	428	456	242	1673
海洋生物医药 Marine Biomedicine	0	0	2	2	2
海洋工程技术研究 Marine Engineering Technology Research	**634**	**281**	**475**	**186**	**794**
海洋化学工程技术 Marine Chemical Engineering Technology	11	8	5	4	41
海洋生物工程技术 Marine Bioengineering Technology	7	5	4	2	8

5-7 续表 continued

行　业 Industry	专利申请受理数 Number of Patent Applications Accepted	#发明专利 Patents for Discoveries	专利授权数 Number of Patents Granted	#发明专利 Patents for Discoveries	拥有发明专利总数 Total Number of Patents for Discoveries
海洋交通运输工程技术 Marine Communications and Transport Engineering Technology	81	29	88	27	104
海洋能源开发技术 Marine Energies Development Technology	5	3	6	1	3
海洋环境工程技术 Marine Environmental Engineering Technology	14	12	33	5	31
河口水利工程技术 Estuarine Water Conservancy Engineering Technology	483	211	291	114	444
其他海洋工程技术 Other Marine Engineering Technologies	33	13	48	33	163
海洋信息服务业 Marine Information Service	**4**	**4**	**5**	**4**	**10**
海洋技术服务业 Marine Technological Service Industry	**1642**	**1176**	**825**	**673**	**2187**
海洋工程管理服务 Marine Engineering Management Service	0	0	0	0	0
其他海洋专业技术服务 Other Marine Professional and Technological Services	1642	1176	825	673	2187

5-8 分行业海洋科研机构R&D情况（2017年）
R&D in the Marine Scientific Research Institutions by Industry (2017)

行　业 Industry	R&D人员 （人） R&D Personnel (person)	R&D经费内部支出 （千元） R&D Internal Expenditure (1000 yuan)	R&D课题数 （项） Number of R&D Projects (item)
合 计 Total	**26056**	**14183198**	**16339**
海洋基础科学研究 Marine Basic Scientific Research	**20493**	**11366707**	**13115**
海洋自然科学 Marine Natural Science	17850	10113603	11651
海洋社会科学 Marine Social Science	293	45066	50
海洋农业科学 Marine Agricultural Science	2298	1192478	1401
海洋生物医药 Marine Biomedicine	52	15560	13
海洋工程技术研究 Marine Engineering Technology Research	**3611**	**1258082**	**2315**
海洋化学工程技术 Marine Chemical Engineering Technology	88	11002	8
海洋生物工程技术 Marine Bioengineering Technology	190	69180	106

5-8 续表 continued

行 业 Industry	R&D人员 （人） R&D Personnel (person)	R&D经费内部支出 （千元） R&D Internal Expenditure (1000 yuan)	R&D课题数 （项） Number of R&D Projects (item)
海洋交通运输工程技术 Marine Communications and Transport Engineering Technology	1020	336205	395
海洋能源开发技术 Marine Energies Development Technology	20	4219	6
海洋环境工程技术 Marine Environmental Engineering Technology	373	196391	112
河口水利工程技术 Estuarine Water Conservancy Engineering Technology	1731	589353	1603
其他海洋工程技术 Other Marine Engineering Technologies	189	51732	85
海洋信息服务业 **Marine Information Service**	**320**	**156141**	**17**
海洋技术服务业 **Marine Technological Service Industry**	**1632**	**1402268**	**892**
海洋工程管理服务 Marine Engineering Management Service	0	0	0
其他海洋专业技术服务 Other Marine Professional and Technological Services	1632	1402268	892

5-9 分地区海洋科研机构及人员情况（2017年）
Marine Scientific Research Institutions and Personnel by Regions (2017)

地　区 Region	机构数（个） Number of Institutions	从业人员（人） Employees (person)
合 计 Total	**159**	**29089**
北　京 Beijing	18	7022
天　津 Tianjin	11	2043
河　北 Hebei	5	532
辽　宁 Liaoning	17	2032
上　海 Shanghai	11	2321
江　苏 Jiangsu	8	1485
浙　江 Zhejiang	19	1990
福　建 Fujian	14	1194
山　东 Shandong	19	3468
广　东 Guangdong	22	4824
广　西 Guangxi	8	436
海　南 Hainan	3	284
其　他 Others	4	1458

5-10 分地区海洋科研机构科技活动人员学历构成（2017年）
Educational Background Composition of the Personnel Engaged in Scientific and Technological Activities in Marine Scientific Research Institutions by Regions (2017)

单位：人 (person)

地　区 Region	科技活动人员 Personnel Engaged in Scientifical Activities	#博士 Doctor	#硕士 Master	#大学生 University Graduate	#大专生 College Graduate
合 计 Total	**25642**	**7811**	**8525**	**6853**	**1541**
北　京 Beijing	6437	2861	1974	1225	237
天　津 Tianjin	1908	236	794	713	125
河　北 Hebei	508	56	132	263	48
辽　宁 Liaoning	1903	307	908	446	141
上　海 Shanghai	2039	515	694	655	109
江　苏 Jiangsu	1342	462	382	326	107
浙　江 Zhejiang	1722	279	677	645	93
福　建 Fujian	1105	193	397	421	53
山　东 Shandong	3019	1037	897	718	215
广　东 Guangdong	3816	1173	1205	934	271
广　西 Guangxi	357	19	85	195	57
海　南 Hainan	232	11	80	102	39
其　他 Others	1254	662	300	210	46

5-11 分地区海洋科研机构科技活动人员职称构成（2017年）
Professional Title Composition of Personel Engaged in Marine Scientific Research Institutions by Regions (2017)

单位：人 (person)

地 区 Region	科技活动人员 Personnel Engaged in Scientifical Activities	#高级职称 Senior Professional Title	#中级职称 Intermediate Professional Title	#初级职称 Junior Professional Title
合 计 Total	**25642**	**11268**	**8823**	**3225**
北 京 Beijing	6437	3495	2055	504
天 津 Tianjin	1908	782	657	294
河 北 Hebei	508	203	104	35
辽 宁 Liaoning	1903	736	693	147
上 海 Shanghai	2039	816	819	277
江 苏 Jiangsu	1342	787	346	157
浙 江 Zhejiang	1722	628	545	334
福 建 Fujian	1105	409	472	165
山 东 Shandong	3019	1189	1223	384
广 东 Guangdong	3816	1446	1210	633
广 西 Guangxi	357	103	125	121
海 南 Hainan	232	65	63	85
其 他 Others	1254	609	511	89

5-12 分地区海洋科研机构经费收入（2017年）
Routine Funds Receipts of Marine Scientific Research Institutions by Regions (2017)

单位：千元 (1000 yuan)

地区 Region	经费收入总额 Fund Total	本年收入合计 Total Annual Income	基本建设中政府投资 Government Investment in the Capital Construction
合计 Total	**26345457**	**24125285**	**2220172**
北京 Beijing	7558230	6915715	642515
天津 Tianjin	1410860	1354671	56189
河北 Hebei	269260	239535	29725
辽宁 Liaoning	1776597	1737454	39143
上海 Shanghai	2933290	2693601	239689
江苏 Jiangsu	1364637	1265761	98876
浙江 Zhejiang	1313638	1279910	33728
福建 Fujian	851995	752799	99196
山东 Shandong	3594084	2865879	728205
广东 Guangdong	3793544	3543107	250437
广西 Guangxi	176142	176142	0
海南 Hainan	107873	107083	790
其他 Others	1195307	1193628	1679

5-13 分地区海洋科研机构科技课题情况（2017年）

Marine Science and Technology Research Projects of the Research Institutions by Regions (2017)

单位：项 (item)

地 区 Region	课题数 Number of Research Projects	基础研究 Basic Research	应用研究 Applied Research	试验发展 Experimental Development	成果应用 Result Application	科技服务 Scientific and Technological Service
合 计 Total	**21257**	**6282**	**6781**	**3276**	**2163**	**2755**
北 京 Beijing	7188	2348	2749	350	700	1041
天 津 Tianjin	835	13	124	262	83	353
河 北 Hebei	84	8	31	19	13	13
辽 宁 Liaoning	492	93	208	35	131	25
上 海 Shanghai	1030	61	308	289	126	246
江 苏 Jiangsu	2711	294	1029	589	597	202
浙 江 Zhejiang	764	179	105	156	100	224
福 建 Fujian	610	286	177	94	15	38
山 东 Shandong	1925	640	714	394	85	92
广 东 Guangdong	3968	1244	1092	870	284	478
广 西 Guangxi	52	15	13	21	2	1
海 南 Hainan	32	2	6	11	13	0
其 他 Others	1566	1099	225	186	14	42

5-14 分地区海洋科研机构科技论著情况（2017年）
Marine Scientific and Technological Theses and Works of the Research Institutions by Regions (2017)

地 区 Region	发表科技论文（篇） Scientific Theses Published (piece)	#国外发表 Published Abroad	出版科技著作（种） Scientific and Technological Works Published (kind)
合 计 Total	**15872**	**7587**	**388**
北 京 Beijing	4609	2621	134
天 津 Tianjin	585	47	31
河 北 Hebei	345	10	57
辽 宁 Liaoning	578	262	5
上 海 Shanghai	954	237	12
江 苏 Jiangsu	1120	303	31
浙 江 Zhejiang	579	172	19
福 建 Fujian	328	166	5
山 东 Shandong	2000	888	43
广 东 Guangdong	3072	1869	38
广 西 Guangxi	58	14	0
海 南 Hainan	89	2	3
其 他 Others	1555	996	10

5-15 分地区海洋科研机构科技专利情况（2017年）

Marine Scientific and Technological Patents of the Research Institutions by Regions (2017)

单位：件 (unit)

地 区 Region	专利申请受理数 Number of Patent Applications Accepted	#发明专利 Patents for Discoveries	专利授权数 Number of Patents Granted	#发明专利 Patents for Discoveries	拥有发明专利总数 Total Number of Patents for Discoveries
合 计 Total	**4779**	**3347**	**3288**	**2242**	**10352**
北 京 Beijing	917	661	748	590	2297
天 津 Tianjin	120	74	143	50	213
河 北 Hebei	3	0	3	1	44
辽 宁 Liaoning	365	278	235	183	580
上 海 Shanghai	308	243	198	151	639
江 苏 Jiangsu	261	158	159	84	462
浙 江 Zhejiang	229	180	169	113	325
福 建 Fujian	47	41	39	29	184
山 东 Shandong	467	319	372	205	1554
广 东 Guangdong	1757	1223	988	725	3437
广 西 Guangxi	10	7	23	8	108
海 南 Hainan	13	11	8	8	49
其 他 Others	282	152	203	95	460

5-16　分地区海洋科研机构R&D情况（2017年）
R&D in the Marine Scientific Research Institutions by Regions（2017）

地　区 Region	R&D人员 （人） R&D Personnel (person)	R&D经费内部支出 （千元） R&D Internal Expenditure (1000 yuan)	R&D课题数 （项） Number of R&D Projects (item)
合 计 Total	**26056**	**14183198**	**16339**
北　京 Beijing	8283	3766120	5447
天　津 Tianjin	1388	740290	399
河　北 Hebei	296	120784	58
辽　宁 Liaoning	1630	1119203	336
上　海 Shanghai	1855	1743318	658
江　苏 Jiangsu	1218	823445	1912
浙　江 Zhejiang	1028	549803	440
福　建 Fujian	840	500465	557
山　东 Shandong	3206	1551766	1748
广　东 Guangdong	3817	2485862	3206
广　西 Guangxi	134	25098	49
海　南 Hainan	62	17654	19
其　他 Others	2299	739390	1510

主要统计指标解释

1. 海洋科研机构 指有明确的研究方向和任务，有一定水平的学术带头人和一定数量、质量的研究人员，有开展研究工作的基本条件，长期有组织地从事海洋研究与开发活动的机构。

2. 从业人员 指由本机构年末直接组织安排工作并支付工资的各类人员总数。包括固定职工、国家有编制的合同制职工、招聘人员和返聘的离退休人员。不包括离退休人员、停薪留职人员。

3. 科技活动人员 指从业人员中的科技管理人员、课题活动人员和科技服务人员。

4. 高级职称 指研究员、副研究员；教授、副教授；高级工程师；高级农艺师；正、副主任医（药、护、技)师；高级实验师；高级统计师；高级经济师；高级会计师；编审(正、副编审)；译审(正、副译审)、高级(主任)记者；正、副研究馆员等。

5. 中级职称 指助理研究员；讲师；工程师；农艺师；主治医(药、护、技)师；实验师；统计师；经济师；会计师；编辑；翻译；记者；馆员等。

6. 初级职称 指研究实习员；助教；助理工程师、技术员；助理农艺师、农业技术员；医(药、护、技)师、医(药、护、技)士；助理实验师、实验员；助理统计师、统计员；助理经济师；助理会计师、会计员；助理编辑、见习编辑；助理翻译；助理记者；助理馆员、管理员等。

7. 政府资金 指由各级政府部门直接拨款或企事业单位利用政府资金委托本机构从事科学技术活动所获得的收入。

8. 基础研究 为获得新知识而进行的独创性研究。其目的是揭示观察到的现象和事实的基本原理和规律，而不以任何特定的实际应用为目的。

9. 应用研究 为获得新的科学技术知识而进行的独创性研究。它主要针对某一特定的实际应用目的。应用研究通常是为了确定基础研究成果或知识的可能的用途，或是为达到某一具体的、预定的实际目的确定新的方法(原理性)或途径。

10. 试验发展 利用从研究或实际经验获得的知识，为生产新的材料、产品和装置，建立新的工艺和系统，以及对已生产或建立的上述各项进行实质性的改进而进行的系统性工作。

11. 成果应用 为解决 R&D 活动阶段产生的新产品、新装置、新工艺、新技术、新方法、新系统和服务等能投入生产或在实际应用中所存在的技术问题而进行的系统性活动。它不具有创新成分。此类活动包括为达到生产目的而进行的定型设计和试制以及为扩大新产品的生产规模和探索新方法、新技术、新工艺等的应用领域而进行的适应性试验。

12. 科技服务 与科学研究与实验发展有关,并有助于科学技术知识的产生、传播和应用的活动。包括为扩大科技成果的使用范围而进行的示范性推广工作；为用户提供科技信息和文献服务的系统性工作；为用户提供可行性报告、技术方案、建议及进行技术论证等技术咨询工作；自然、生物现象的日常观测、监测，资源的考查和勘探；有关社会、人文、经济现象的通用资料的收集，如统计、市场调查等以及这些资料的常规分析与整理；为社会和公众提供的测试、标准化、计量、计算、质量控制和专利服务，不包括工商企业为进行正常生产而开展的上述活动。

13. 科技论文 在全国性学报或学术刊物上、省部属大专院校对外正式发行的学报或学术刊物上发表的论文以及向国外发表的论文。

14. 科技著作 经过正式出版部门编印出版的科技专著、大专院校教科书、科普著作。
15. 专利申请受理数 当年本单位向专利管理部门提出申请并被受理的职务专利申请件数。
16. 专利授权数 当年由专利管理部门授予本单位专利权的职务专利件数。

Explanatory Notes on Main Statistical Indicators

1. Marine Scientific Research Institution refers to the institution which has definite research orientations and tasks, high-level academic leading personnel and fair-sized, qualified research personnel, and basic conditions for research work and which is engaged for a long time in the marine research and development activities in an organized way.
2. Employees refers to the total number of personnel of various kinds employed and paid by the institution at the end of the year, including fixed employees, contract workers of staff belonging to the state authorized staff , recruited personnel, reemployed retired personnel, but not including the retired and the personnel on leave with pay suspension.
3. Personnel Engaged in Scientific and Technological Activities refers to the personnel for scientific and technological management, personnel engaged in the activities of research topics and scientific and technological service personnel.
4. Senior Professional refers to research scientist, associate research scientist; professor, associate professor; senior engineer; senior agronomist; professor-rank and associate professor-rank doctor (pharmacists, nurses and technicians); senior laboratory technician; senior statisticians; senior economic engineer; chief accountant; senior editor (professor and associate professor ranks); senior translator (professor and associate professor ranks); senior journalist; research librarian (professor and associate professor ranks), etc.
5. Intermediate Professional Title refers to assistant research scientist; lecturer; engineer; agronomist; lecturer-rank doctor (pharmacist, nurse and technician); laboratory technician; lecturer-rank statistician; economic engineer; accountant; editor; translator; journalist; librarian, etc.
6. Junior Professional Title refers to research assistant; assistant; assistant engineer, technician, assistant agronomist, agricultural technician; assistant-rank doctor (pharmacist, nurse and technician); assistant laboratory technician; assistant statistician; assistant economic engineer, assistant accountant; assistant editor, editor on probation; assistant translator; assistant journalist; assistant research librarian, librarian ,etc.
7. Funds from Government refers to the direct appropriation by the government departments at all levels or the earnings from conducting scientific and technological activities entrusted to the institution by enterprises and institutions by using the funds from government.
8. Basic Research refers to the original research to acquire new knowledge. It is aimed at revealing the basic principles and laws of the phenomena and facts observed, but not at any specific practical

applications.

9. Applied Research refers to the original research to acquire new scientific and technological knowledge. It mainly serves the purpose of a particular practical application. The purpose of applied research is usually to define the potential uses of the research finds or knowledge obtained from basic research or to identify new methods (principles) or ways to reach a specific and predetermined goal.

10. Experimental Development refers to the systematic work carried out to establish new technologies and systems for producing new materials, products and equipment by using the knowledge obtained from research or practical experience, or to make substantial improvement of the above-mentioned which have been produced or established.

11. Results Application refers to the systematic activities carried out to solve the technical problems that might crop up in the production or practical application of the new products, devices, technologies, techniques, methods, systems and service occurring in the course of R & D activities. They do not bring forth new ideas. Such activities include the finalizing design and trial-production for the purpose of production as well as the adaptive tests to expand the production scale of new products and the application areas of new methods, techniques and technologies.

12. Scientific and Technological Service refers to the activities that are associated with the scientific research and experimental development, and contribute to the generation, dissemination and application of scientific and technological knowledge, which include the demonstrative work of popularization to enlarge the use scope of scientific and technological achievements; the systematic work of providing the users with scientific and technological information and literature service; the technical consultation work of providing users with feasibility reports, technical schemes and recommendations and carrying out technical demonstration; routine observation and monitoring of natural and biological phenomena, and survey and exploration of resources; collection of universal data on the appropriate social, cultural and economic phenomena, such as statistics and market survey, as well as the routine analysis and sorting-out of these data; the provision for the society and the public of such service as testing, standardization, metering computation, quality control and patent service, but not including the type of the above-mentioned activities carried out by industrial and commercial enterprises for the purpose of normal production.

13. Scientific Treatises refers to the theses published in the national journals or academic publications, those officially issued journals or academic publications by universities and colleges under provinces or ministries as well as theses published abroad.

14. Scientific and Technological Works refers to the scientific and technological monographs, textbooks for universities and colleges and popular science books published by the official publishing houses.

15. Number of Patent Applications Accepted refers to the number of professional patent applications of the unit to the patent administrative department and accepted by it in the year.

16. Number of Patents Granted refers to the number of the professional patents granted to the unit by the patent administrative department in the year.

6

海 洋 教 育

Marine Education

6-1 全国各海洋专业博士研究生情况（2017年）
Doctoral Students from Marine Specialities (2017)

专 业 Speciality	专业点数（个） Number of Speciality Agencies	学生数（人） Number of Students (person)			
		毕业生 Graduates	招 生 Entrants	在校生 Enrollment	预计毕业生数 Estimated Graduates of Next Year
合 计 Total	**138**	**733**	**1113**	**4984**	**2454**
物理海洋学 Physical Oceanography	6	56	67	298	152
海洋化学 Marine Chemistry	7	23	46	182	101
海洋生物学 Marine Biology	8	70	99	364	122
海洋地质 Marine Geology	8	47	65	262	138
海洋科学学科 Marine Science Subjects	6	89	127	567	276
水生生物学 Hydrobiology	19	85	86	334	152
水文学及水资源 Hydrology and Water Resource	18	95	142	688	347
港口海岸及近海工程 Coastal Harbour and Offshore Engineering	10	32	59	370	221

6-1 续表 continued

专 业 Speciality	专业点数（个） Number of Speciality Agencies	学生数（人） Number of Students (person)			
		毕业生 Graduates	招 生 Entrants	在校生 Enrollment	预计毕业生数 Estimated Graduates of Next Year
船舶与海洋结构物设计制造 Ships and Marine Structures Design and Manufacture	11	49	82	429	247
轮机工程 Turbine Engineering	8	32	54	312	141
水声工程 Hydroacoustic Engineering	6	22	52	227	107
船舶与海洋工程学科 Ship and Ocean Engineering Subjects	5	28	63	330	130
水产品加工及贮藏工程 Aquatic Products Processing and Storing Engineering	5	6	3	30	22
水产养殖 Aquaculture	7	54	88	277	135
捕捞学 Science of Fishing	2	3	8	25	15
渔业资源 Fishery Resource	5	15	28	120	64
水产学科 Fishery Subjects	6	27	44	168	83
航空、航天与航海医学 Aeronautical, Aerospace and Nautical Medicine	1	0	0	1	1

6-2 全国各海洋专业硕士研究生情况（2017年）
Postgraduate Students from Marine Specialities (2017)

专 业 Speciality	专业点数（个） Number of Speciality Agencies	学生数（人） Number of Students (person)			
		毕业生 Graduates	招 生 Entrants	在校生 Enrollment	预计毕业生数 Estimated Graduates of Next Year
合 计 **Total**	**312**	**3102**	**3741**	**10472**	**3326**
物理海洋学 Physical Oceanography	14	96	168	483	145
海洋化学 Marine Chemistry	16	101	121	335	96
海洋生物学 Marine Biology	21	251	259	800	278
海洋地质 Marine Geology	12	118	130	392	123
海洋科学学科 Marine Science Subjects	18	324	494	1178	333
水生生物学 Hydrobiology	43	177	217	635	205
水文学及水资源 Hydrology and Water Resource	45	399	388	1106	352
港口海岸及近海工程 Coastal Harbour and Offshore Engineering	20	256	218	680	239

6-2 续表 continued

专 业 Speciality	专业点数（个） Number of Speciality Agencies	学生数（人） Number of Students (person)			
		毕业生 Graduates	招 生 Entrants	在校生 Enrollment	预计毕业生数 Estimated Graduates of Next Year
船舶与海洋结构物设计制造 Ships and Marine Structures Design and Manufacture	16	258	271	845	298
轮机工程 Turbine Engineering	11	219	261	755	261
水声工程 Hydroacoustic Engineering	8	121	127	350	113
船舶与海洋工程学科 Ship and Ocean Engineering Subjects	14	227	265	726	245
水产品加工及贮藏工程 Aquatic Products Processing and Storing Engineering	24	64	50	155	58
水产养殖 Aquaculture	26	355	502	1317	366
捕捞学 Science of Fishing	4	17	22	60	16
渔业资源 Fishery Resource	11	63	103	273	80
水产学科 Fishery Subjects	9	56	145	382	118

6-3 全国普通高等教育各海洋专业本科学生情况（2017年）
Undergraduates from Marine Specialities in the National Ordinary Higher Education (2017)

专 业 Speciality	专业点数（个） Number of Speciality Agencies	学生数（人） Number of Students (person)			
		毕业生 Graduates	招 生 Entrants	在校生 Enrollment	预计毕业生数 Estimated Graduates of Next Year
合 计 Total	**288**	**18508**	**22005**	**84395**	**20908**
海洋科学 Marine Science	28	1005	1456	5216	1262
海洋技术(注：可授理学或工学学士学位) Marine Technology (Note: It may confer bachelor's degrees in science and engineering)	21	632	643	3100	743
海洋资源与环境 Marine Living Resources and Environment	11	322	514	1773	432
海洋科学类专业 New Specialties Under the Category of Marine Sciences	10	105	1065	1513	95
港口航道与海岸工程 Harbour Channel and Coastal Engineering	34	1833	1705	7518	2080
航海技术 Nautical Technology	8	380	436	1685	455
轮机工程 Turbine Engineering	18	2137	2504	10308	2687
船舶与海洋工程 Ship and Marine Engineering	24	2797	3016	12319	3299
海洋工程与技术 Ocean Engineering and Technology	33	5088	5570	21434	5188
海洋资源开发技术 Marine Resources Exploitation Technology	5	88	118	616	107
海洋工程类专业 New Speciality of Marine Engineering	12	255	410	1284	256
水产养殖学 Aquaculture	55	3022	2943	12467	3332
海洋渔业科学与技术 Science and Technology of Marine Fishery	10	439	449	2124	493
水族科学与技术 Science and Technology of Aquatic Animals	11	291	286	1474	389
水产类专业 Fishery-type Specialities	3	0	526	598	0
海事管理 Maritime Affairs Management	5	114	364	966	90

6-4 全国普通高等教育各海洋专业专科学生情况（2017年）
Students from Marine Specialities of the Colleges for Professional Training in the National Ordinary Higher Education (2017)

专 业 Speciality	专业点数（个） Number of Speciality Agencies	学生数（人） Number of Students (person)			
		毕业生 Graduates	招 生 Entrants	在校生 Enrollment	预计毕业生数 Estimated Graduates of Next Year
合 计 Total	**943**	**56144**	**46462**	**155906**	**56160**
水产养殖技术 Aquaculture Technology	35	1148	1137	3692	1144
海洋渔业技术 Marine Fishery Technology	1	0	0	2	0
水族科学与技术 Aquarium Science and Technology	2	14	48	95	6
水生动物医学 Aquatic Animal Medicine	2	0	17	31	0
渔业经济管理 Fishery Economic Management	1	0	29	29	0
渔业类专业 Fishery	1	42	0	0	0
钻井技术 Drilling Technology	10	641	195	832	399
油气开采技术 Oil and Gas Exploitation Technology	16	1164	300	1229	553
油气储运技术 Oil and Gas Storage and Transportation Technology	25	1514	857	3195	1295
油气地质勘探技术 Oil and Gas Geological Exploration Technology	9	396	185	1197	627
油田化学应用技术 Oilfield Chemical Technology	6	269	193	687	211
石油工程技术 Petroleum Engineering Technology	14	1493	598	2932	1335
石油与天然气类专业 Petroleum and Natural Gas	2	166	0	1	0
水文与水资源工程 Hydrology and Water Resources Engineering	12	251	416	1044	265

6-4 续表1 continued

专 业 Speciality	专业点数（个） Number of Speciality Agencies	学生数（人） Number of Students (person)			
		毕业生 Graduates	招 生 Entrants	在校生 Enrollment	预计毕业生数 Estimated Graduates of Next Year
水文测报技术 Hydrological Forecasting Technology	2	38	2	16	6
水政水资源管理 Water Administration and Water Resources Management	5	162	243	719	291
水文水资源类专业 Hydrology and Water Resources	2	80	0	105	104
水利工程 Water Conservancy Projects	35	3709	3374	11152	3960
水利水电工程技术 Water Conservancy and Hydropower Engineering Technology	17	1010	1503	3766	885
水利水电工程管理 Water Conservancy and Hydropower Project Management	18	1201	1419	4228	1342
水利水电建筑工程 Water Conservancy and Hydropower Construction	51	6429	4624	17045	6222
水务管理 Water Affairs Management	4	87	226	523	100
水利工程与管理类专业 Water Conservancy and Management	2	272	535	1353	357
水电站动力设备 Power Equipment of Hydropower Station	9	228	102	666	350
水电站电气设备 Electrical Equipment of Hydropower Station	1	83	0	48	48
水电站运行与管理 Operation and Management of Hydropower Station	4	50	215	396	0
水利机电设备运行与管理 Operation and Management of Electrical and Mechanical Equipment in Water Conservancy	3	4	62	158	38

6-4 续表2 continued

专 业 Speciality	专业点数（个） Number of Speciality Agencies	学生数（人） Number of Students (person)			
		毕业生 Graduates	招 生 Entrants	在校生 Enrollment	预计毕业生数 Estimated Graduates of Next Year
水利水电设备类专业 Water Conservancy and Hydroelectric Equipment	1	45	45	139	44
水土保持技术 Soil and Water Conservation Technology	15	321	144	738	286
水环境监测与治理 Water Environmental Monitoring and Protection	6	172	140	319	107
水土保持与水环境类专业 Soil Conservation and Water Environment	1	29	16	44	28
船舶工程技术 Ship Engineering Technology	36	3261	1786	7385	3187
船舶机械工程技术 Ship Mechanical Engineering Technology	11	511	301	1376	579
船舶电气工程技术 Ship Electrical Engineering Technology	14	705	599	2076	854
船舶舾装工程技术 Ship Equipment and Installations	3	90	99	255	61
船舶涂装工程技术 Engineering Technology of Ship Painting	1	59	40	134	61
游艇设计与制造 Yacht Design and Manufacturing	7	251	226	677	244
海洋工程技术 Marine Engineering Technology	6	335	159	673	354
船舶通信与导航 Ship Communication and Navigation	3	124	157	418	149
船舶动力工程技术 Ship Power Engineering Technology	6	373	316	1166	377

6-4 续表3 continued

专 业 Speciality	专业点数（个） Number of Speciality Agencies	学生数（人） Number of Students (person)			
		毕业生 Graduates	招 生 Entrants	在校生 Enrollment	预计毕业生数 Estimated Graduates of Next Year
航海技术 Nautical Technology	50	4735	4866	14967	5036
国际邮轮乘务管理 International Cruise Ship Management	85	2332	4886	12047	3352
船舶电子电气技术 Electronic and Electrical Technology of Ships	20	663	578	2194	671
船舶检验 Ship Inspection	8	180	149	605	233
港口机械与自动控制 Port Machinery and Automatic Control	21	1084	946	2945	1141
港口电气技术 Port Electrical Technology	4	218	135	570	233
港口与航道工程技术 Port and Waterway Engineering Technology	12	523	332	1278	562
港口与航运管理 Harbour and Shipping Management	47	3565	2494	8596	3224
港口物流管理 Port Logistics Management	15	426	631	1840	514
轮机工程技术 Turbine Engineering Technology	51	3540	2805	9626	3535
水路运输与海事管理 Waterway Transportation and Maritime Management	16	697	748	2067	688
集装箱运输管理 Container Transportation Management	12	517	398	1370	474
水上运输类专业 Water Transportation	5	140	70	269	85
报关与国际货运 Customs Declaration and International Freight Transport	198	10797	7116	26991	10543

6-5 全国成人高等教育各海洋专业本科学生情况（2017年）
Undergraduates from Marine Specialities in the National Adult Higher Education (2017)

专业 Speciality	专业点数（个） Number of Speciality Agencies	学生数（人） Number of Students (person)			
		毕业生 Graduates	招生 Entrants	在校生 Enrollment	预计毕业生数 Estimated Graduates of Next Year
合计 Total	**59**	**3153**	**2085**	**6282**	**3161**
海洋科学 Marine Science	1	27	0	3	3
港口航道与海岸工程 Harbour Channel and Coastal Engineering	4	42	19	65	39
航海技术 Navigation Technology	9	213	236	600	236
轮机工程 Turbine Engineering	11	260	338	838	301
船舶与海洋工程 Ship and Marine Engineering	12	2348	1310	4194	2292
水产养殖学 Aquaculture	22	263	182	582	290

6-6 全国成人高等教育各海洋专业专科学生情况（2017年）
Students from Marine Specialities of the Colleges for Professional Training in the National Adult Higher Education (2017)

专 业 Speciality	专业点数（个） Number of Speciality Agencies	学生数（人） Number of Students (person)			
		毕业生 Graduates	招 生 Entrants	在校生 Enrollment	预计毕业生数 Estimated Graduates of Next Year
合 计 Total	**338**	**12067**	**6203**	**16719**	**8139**
水产养殖技术 Aquaculture Technology	25	736	427	1343	529
渔业类专业 Fishery	1	1	6	8	1
钻井技术 Drilling Technology	5	28	23	38	14
石油与天然气类专业 Oil and Gas	2	9	8	12	4
油气开采技术 Oil and Gas Exploitation Technology	12	574	76	235	140
油气储运技术 Oil and Gas Storage and Transportation Technology	13	270	68	292	187
油气地质勘探技术 Oil and Gas Geology Exploration Technology	6	600	14	58	44
油田化学应用技术 Oilfield Chemical Technology	1	0	0	1	1
石油工程技术 Petroleum Engineering Technology	14	1496	1065	2038	865
水文与水资源工程 Hydrology and Water Resources Engineering	4	12	11	14	3
水文测报技术 Hydrological Forecasting Technology	1	1	0	0	0
水政水资源管理 Water Administration and Water Resources Management	4	39	6	30	23

6-6 续表1 continued

专 业 Speciality	专业点数（个） Number of Speciality Agencies	学生数（人） Number of Students (person)			
		毕业生 Graduates	招 生 Entrants	在校生 Enrollment	预计毕业生数 Estimated Graduates of Next Year
水文水资源类专业 Hydrology and Water Resources	3	15	38	64	11
水利工程 Water Conservancy Projects	20	529	481	783	279
水利水电工程技术 Water Conservancy and Hydropower Engineering Technology	4	7	17	53	19
水利水电工程管理 Water Conservancy and Hydropower Project Management	35	1184	811	2175	919
水利水电建筑工程 Water Conservancy and Hydropower Construction	30	2139	1140	2617	1429
水利工程与管理类专业 Water Conservancy and Management	7	613	328	846	464
水电站动力设备 Power Equipment of Hydropower Station	3	1	4	5	1
水利水电设备类专业 Water Conservancy and Hydroelectric Equipment	2	27	57	159	102
水土保持技术 Soil and Water Conservation	5	74	32	55	19

6-6 续表2 continued

专 业 Speciality	专业点数（个） Number of Speciality Agencies	学生数（人） Number of Students (person)			
		毕业生 Graduates	招 生 Entrants	在校生 Enrollment	预计毕业生数 Estimated Graduates of Next Year
船舶工程技术 Ship Engineering	21	844	365	982	409
船舶机械工程技术 Ship Mechanical Engineering	2	14	0	2	2
航海技术 Nautical Technology	33	1290	522	2281	1338
国际邮轮乘务管理 International Cruise Crew Management	6	118	111	285	147
船舶电子电气技术 Electronic and Electrical Technology of Ships	3	35	11	139	64
船舶检验 Ship Inspection	1	3	1	1	0
港口机械与自动控制 Port Machinery and Automatic Control	1	51	55	159	55
港口与航运管理 Harbour and Shipping Management	10	209	92	273	55
港口物流管理 Port Logistics Management	2	13	1	2	1
轮机工程技术 Turbine Engineering	31	939	378	1466	823
水路运输与海事管理 Waterway Transportation and Maritime Management	2	13	0	78	30
水上运输类专业 Water Transportation	3	26	0	80	80
报关与国际货运 Customs Declaration and International Freight Transport	26	157	55	145	81

6-7 全国中等职业教育各海洋专业学生情况（2017年）
Students from Marine Specialities in the National Secondary Vocational Education (2017)

专 业 Speciality	专业点数（个） Number of Speciality Agencies	学生数（人） Number of Students (person)			
		毕业生 Graduates	招 生 Entrants	在校生 Enrollment	预计毕业生数 Estimated Graduates of Next Year
合 计 Total	**253**	**15363**	**11995**	**32158**	**12253**
海水生态养殖 Seawater Ecological Cultivation	12	832	243	843	323
航海捕捞 Sea Fishing	4	65	1255	1693	1348
农林牧渔类新专业 New Specialities of Agriculture, Forestry, Animal Husbandry and Fishery	36	2127	2294	5434	1828
水文与水资源勘测 Hydrological and Water Resources Survey	3	126	0	87	45
风电场机电设备运行与维护 Operation and Maintenance of Electromechanical Equipment in the Wind Power Station	23	1198	957	2882	1022
船舶制造与修理 Ships Building and Repair	28	2588	1527	5570	1690
船舶机械装置安装与维修 Installation and Maintenance of Ships' Mechanical Equipment	3	415	567	1618	564

注：此表不包含技工学校相关数据。

Note: Data related to Technical Schools are not included in the table.

6-7 续表 continued

专 业 Speciality	专业点数（个） Number of Speciality Agencies	学生数（人） Number of Students (person)			
		毕业生 Graduates	招 生 Entrants	在校生 Enrollment	预计毕业生数 Estimated Graduates of Next Year
船舶驾驶 Ship Piloting	47	3272	2265	6210	2540
轮机管理 Engines Management	43	2918	1362	3300	1444
船舶水手与机工 Ship Sailors and Mechanics	12	367	522	1256	341
船舶电气技术 Ship Electric Technology	11	529	85	493	269
外轮理货 Foreign Ships Freight Forwarding	7	132	35	393	143
船舶检验 Ships Inspection	2	30	0	27	0
港口机械运行与维护 Operation and Maintenance of Harbour Machinery	20	736	846	2272	678
工程潜水 Engineering Diving	2	28	37	80	18

6-8 分地区各海洋专业博士研究生情况（2017年）
Doctoral Students in Marine Specialities by Regions (2017)

地 区 Region	专业点数（个） Number of Speciality Agencies	学生数（人） Number of Students (person)			
		毕业生 Graduates	招 生 Entrants	在校生 Enrollment	预计毕业生数 Estimated Graduates of Next Year
合 计 Total	**138**	**733**	**1113**	**4984**	**2454**
北 京 Beijing	13	165	213	740	297
天 津 Tianjin	3	3	8	35	11
辽 宁 Liaoning	8	28	74	390	191
上 海 Shanghai	20	59	120	528	233
江 苏 Jiangsu	11	62	98	535	268
浙 江 Zhejiang	5	25	37	165	80
福 建 Fujian	8	35	71	275	84
山 东 Shandong	13	167	189	979	597
广 东 Guangdong	12	26	51	175	70
广 西 Guangxi	1	1	1	6	3
海 南 Hainan	1	3	6	11	1
其 他 Others	43	159	245	1145	619

6-9 分地区各海洋专业硕士研究生情况（2017年）
Postgraduate Students in Marine Specialities by Regions (2017)

地　区 Region	专业点数 （个） Number of Speciality Agencies	学生数（人） Number of Students (person)			
		毕业生 Graduates	招　生 Entrants	在校生 Enrollment	预计毕业生数 Estimated Graduates of Next Year
合　计 Total	**312**	**3102**	**3741**	**10472**	**3326**
北　京 Beijing	20	198	261	760	251
天　津 Tianjin	10	79	102	265	80
河　北 Hebei	5	12	12	35	15
辽　宁 Liaoning	24	315	380	1066	339
上　海 Shanghai	28	375	476	1231	373
江　苏 Jiangsu	27	346	396	1125	369
浙　江 Zhejiang	25	255	297	882	263
福　建 Fujian	18	159	200	536	172
山　东 Shandong	31	296	410	1178	370
广　东 Guangdong	17	144	187	497	145
广　西 Guangxi	4	14	30	66	17
海　南 Hainan	4	19	25	71	23
其　他 Others	99	890	965	2760	909

6-10 分地区普通高等教育各海洋专业本科学生情况（2017年）
Undergraduates in the Marine Specialities of Ordinary Higher Education by Regions (2017)

地 区 Region	专业点数（个）Number of Speciality Agencies	学生数（人）Number of Students (person)			
		毕业生 Graduates	招 生 Entrants	在校生 Enrollment	预计毕业生数 Estimated Graduates of Next Year
合 计 Total	**288**	**18508**	**22005**	**84395**	**20908**
北 京 Beijing	3	142	188	655	150
天 津 Tianjin	14	946	1060	4084	974
河 北 Hebei	11	388	334	1667	443
辽 宁 Liaoning	26	2159	2592	10166	2548
上 海 Shanghai	16	1160	1334	5358	1434
江 苏 Jiangsu	33	1638	2184	7314	1574
浙 江 Zhejiang	31	1184	1252	5457	1423
福 建 Fujian	16	1428	1621	6644	1779
山 东 Shandong	34	2466	2873	10820	2490
广 东 Guangdong	22	1487	2206	7985	2157
广 西 Guangxi	9	280	523	1797	258
海 南 Hainan	7	139	489	1342	119
其 他 Others	66	5091	5349	21106	5559

6-11 分地区普通高等教育各海洋专业专科学生情况（2017年）
Students from the Marine Specialities Colleges for Professional Training in the Ordinary Higher Education by Regions (2017)

地 区 Region	专业点数（个） Number of Speciality Agencies	学生数（人） Number of Students (person)			
		毕业生 Graduates	招 生 Entrants	在校生 Enrollment	预计毕业生数 Estimated Graduates of Next Year
合 计 Total	**943**	**56144**	**46462**	**155906**	**56160**
北 京 Beijing	2	76	128	244	48
天 津 Tianjin	28	3321	2710	8764	3258
河 北 Hebei	53	1817	1477	4870	1828
辽 宁 Liaoning	61	4233	3491	12250	4546
上 海 Shanghai	24	1438	856	3162	1139
江 苏 Jiangsu	80	4668	3285	13061	5035
浙 江 Zhejiang	41	3254	2405	8753	3353
福 建 Fujian	48	2104	2624	7706	2489
山 东 Shandong	108	8002	6342	22192	8396
广 东 Guangdong	39	2268	2493	7139	2270
广 西 Guangxi	47	1759	1603	4947	1747
海 南 Hainan	13	1527	547	1594	581
其 他 Others	399	21677	18501	61224	21470

6-12 分地区成人高等教育各海洋专业本科学生情况（2017年）
Students from Marine Specialities in the Adult Higher Education by Regions (2017)

地 区 Region	专业点数（个） Number of Speciality Agencies	学生数（人） Number of Students (person)			
		毕业生 Graduates	招 生 Entrants	在校生 Enrollment	预计毕业生数 Estimated Graduates of Next Year
合 计 Total	**59**	**3153**	**2085**	**6282**	**3161**
天 津 Tianjin	2	57	4	49	45
河 北 Hebei	1	4	0	24	8
辽 宁 Liaoning	5	255	265	602	337
上 海 Shanghai	4	154	197	462	152
江 苏 Jiangsu	5	1753	1106	3221	1750
浙 江 Zhejiang	4	109	48	214	46
福 建 Fujian					
山 东 Shandong	11	179	147	263	116
广 东 Guangdong	7	149	70	436	200
广 西 Guangxi	2	2	6	12	6
其 他 Others	18	491	242	999	501

6-13 分地区成人高等教育各海洋专业专科学生情况（2017年）
Students from Marine Specialities of the Colleges for Professional Training in the Adult Higher Education by Regions (2017)

地　区 Region	专业点数（个） Number of Speciality Agencies	学生数（人） Number of Students (person)			
		毕业生 Graduates	招生 Entrants	在校生 Enrollment	预计毕业生数 Estimated Graduates of Next Year
合　计 Total	**338**	**12067**	**6203**	**16719**	**8139**
天　津 Tianjin	6	90	16	75	59
河　北 Hebei	11	235	199	585	241
辽　宁 Liaoning	31	800	519	1792	995
上　海 Shanghai	10	432	246	1168	382
江　苏 Jiangsu	32	1552	802	2108	1101
浙　江 Zhejiang	13	149	58	519	461
福　建 Fujian	6	533	294	995	397
山　东 Shandong	33	839	343	705	360
广　东 Guangdong	17	271	136	495	192
广　西 Guangxi	5	30	8	44	36
海　南 Hainan	2	1	3	8	5
其　他 Others	172	7135	3579	8225	3910

6-14 分地区中等职业教育各海洋专业学生情况（2017年）
Students from Marine Specialities in the Secondary Vocational Education by Regions (2017)

地 区 Region	专业点数（个） Number of Speciality Agencies	学生数（人） Number of Students (person)			
		毕业生 Graduates	招 生 Entrants	在校生 Enrollment	预计毕业生数 Estimated Graduates of Next Year
合 计 Total	**253**	**15363**	**11995**	**32158**	**12253**
天 津 Tianjin	4	710	614	1331	636
河 北 Hebei	17	1204	976	2296	593
辽 宁 Liaoning	22	444	393	1405	507
上 海 Shanghai	14	1003	1169	3526	1118
江 苏 Jiangsu	24	947	1080	2677	837
浙 江 Zhejiang	14	624	451	1417	549
福 建 Fujian	27	4484	2166	3961	3006
山 东 Shandong	42	1366	1045	3305	1201
广 东 Guangdong	11	332	443	1280	328
广 西 Guangxi	8	547	1014	2753	855
海 南 Hainan	2	80	53	140	21
其 他 Others	68	3622	2591	8067	2602

注：此表不包含技工学校相关数据。

Note: Data related to Technical Schools are not included in the table.

6-15 分地区开设海洋专业高等学校教职工数（2017年）
Number of Teaching and Administrative Staff in the Universities and Colleges Offering Marine Specialities by Regions (2017)

地　区 Region	学校（机构）数 （个） Number of Colleges (Institutions) (unit)	教职工数 （人） Number of Teaching and Administrative Staff (person)	专任教师数 （人） Number of Full-Time Teachers (person)
合　计 Total	**595**	**778009**	**497589**
北　京 Beijing	12	41983	22171
天　津 Tianjin	15	18741	12879
河　北 Hebei	32	31854	20643
辽　宁 Liaoning	24	22782	14859
上　海 Shanghai	20	31687	17402
江　苏 Jiangsu	53	76457	50307
浙　江 Zhejiang	25	33255	20558
福　建 Fujian	23	27739	16993
山　东 Shandong	50	63521	43375
广　东 Guangdong	31	56173	33431
广　西 Guangxi	23	21690	14887
海　南 Hainan	9	7369	4660
其　他 Others	278	344758	225424

主要统计指标解释

海洋专业 指高等教育和中等职业教育所设的与海洋有关的专业。

Explanatory Notes on Main Statistical Indicators

Marine Speciality refers to the marine-related speciality in the higher education and the secondary vocational education.

7
海洋环境保护
Marine Environmental Protection

7-1 海区海水水质评价结果（2017年）
Seawater Quality Assessment Results by Sea Area (2017)

海区 Sea Area	季节 Season	合计 （平方千米） **Total** **(km²)**	第二类水质海域面积 （平方千米） Area of the Second Grade Sea Waters (km²)	第三类水质海域面积 （平方千米） Area of the Third Grade Sea Waters (km²)
全海域 Total Area	夏季 Summer	**130330**	49830	28540
	秋季 Autumn	**188000**	71970	38440
渤海 Bohai Sea	夏季 Summer	**18740**	8940	3970
	秋季 Autumn	**32480**	15710	8300
黄海 Yellow Sea	夏季 Summer	**28220**	17280	7090
	秋季 Autumn	**45240**	20980	10980
东海 East China Sea	夏季 Summer	**60480**	17610	9260
	秋季 Autumn	**80000**	23380	10260
南海 South China Sea	夏季 Summer	**22890**	6000	8220
	秋季 Autumn	**30280**	11900	8900

7-1 续表 continued

海 区 Sea Area	季 节 Season	第四类水质海域面积 （平方千米） Area of the Fourth Grade Sea Waters (km^2)	劣于第四类水质海域面积 （平方千米） Sea Area of the Sea Waters Inferior to the Fourth Grade (km^2)	首要超标污染物 Prime Pollutants Exceeding the Set Standard
全海域 Total Area	夏季 Summer	18240	33720	无机氮、活性磷酸盐、石油类 Inorganic Nitrogen, Active Phosphate, Petroleum
	秋季 Autumn	30280	47310	
渤 海 Bohai Sea	夏季 Summer	2120	3710	
	秋季 Autumn	4780	3690	
黄 海 Yellow Sea	夏季 Summer	2610	1240	
	秋季 Autumn	9440	3840	
东 海 East China Sea	夏季 Summer	11400	22210	
	秋季 Autumn	11850	34510	
南 海 South China Sea	夏季 Summer	2110	6560	
	秋季 Autumn	4210	5270	

7-2 海区废弃物海洋倾倒情况（2017年）
Ocean Dumping of Wastes by Sea Area (2017)

单位：万立方米 (10000 m^3)

海 区 Sea Area	海洋废弃物 Ocean Dumping Wastes
合 计 Total	**15771**
渤黄海 Bohai Sea and Yellow Sea	2273
东 海 East China Sea	10929
南 海 South China Sea	2569

7-3 海区海洋石油勘探开发污染物排放入海情况（2017年）
Discharge of Pollutants into the Sea from Offshore Oil Exploration and Exploitation (2017)

海　区 Sea Area	生产污水（万立方米） Production Sewage (10000 m^3)	泥浆（立方米） Sludge (m^3)	钻屑（立方米） Debris from Drilling (m^3)	机舱污水（立方米） Sewage from Engineroom (m^3)	食品废弃物（吨） Food Wastes (t)	生活污水（立方米） Domestic Sewage (m^3)
合　计 Total	**18917.3**	**34117**	**44948**	**1770.4**	**501.9**	**73**
渤黄海 Bohai Sea and Yellow Sea	759.7	5908	25962		3.6	38
东　海 East China Sea	102.9	303	682	4.5		5
南　海 South China Sea	18054.7	27906	18304	1765.9	498.3	30

7-4 沿海区域海洋类型保护区建设情况（2017年）
Construction of Marine-Type Reserves in Coastal Regions (2017)

地　区 Region	保护区面积（平方千米）Area of Nature Reserves (km^2)	保护区数量（个）Number of Nature Reserves (unit)	按保护级别分（个）by Level of Protection		其中：自然保护区按保护类型分（个）Including: Nature Reserves by Type of Protection			
			国家级 National	地方级 Provincial	海洋和海岸自然生态系统 Marine and Coastal Natural Ecosystems	海洋自然遗迹和非生物资源 Marine Natural Relics and Non-living Resources	海洋生物物种 Marine Species	其他 Others
合　计 Total	**97697**	**195**	**89**	**106**	**60**	**8**	**42**	**5**
环渤海地区 Round-the-Bohai Sea Region	14502	65	49	16	36	2	6	0
长江三角洲地区 Yangtze River Delta Region	60984	26	14	12	4	1	1	5
海峡西岸地区 Region on the West Side of the Taiwan Straits	1828	50	11	39	5	2	5	0
珠江三角洲地区 Zhujiang River Delta	3387	50	11	39	15	0	28	0
环北部湾地区 Round-the-Beibu Gulf Region	16996	4	4	0	0	0	2	0

注：环渤海地区保护区数量不包含辽宁省数据。
Note: The number of nature reserves in Round-the-Bohai Sea Region doesn't include those in Liaoning Province.

7-5 沿海地区海洋类型保护区建设情况（2017年）
Construction of Marine-Type Reserves (2017)

地　区 Region	保护区面积（平方千米）Area of Nature Reserves (km^2)	保护区数量（个）Number of Nature Reserves (unit)	按保护级别分（个）by Level of Protection (unit)		其中：自然保护区按保护类型分（个）Including: Nature Reserves by Type of Protection (unit)			
			国家级 National	地方级 Provincial	海洋和海岸自然生态系统 Marine and Coastal Natural Ecosystems	海洋自然遗迹和非生物资源 Marine Natural Relics and Non-living Resources	海洋生物物种 Marine Species	其他 Others
合　计 Total	**123026**	**184**	**93**	**91**	**60**	**25**	**43**	**5**
天　津 Tianjin	393	2	2	0	1	0	0	0
河　北 Hebei	344	3	1	2	2	1	0	0
辽　宁 Liaoning	7342							
上　海 Shanghai	941	4	2	2	1	0	0	3
江　苏 Jiangsu	57701	3	3	0	0	0	0	0
浙　江 Zhejiang	2342	19	9	10	3	1	1	2
福　建 Fujian	1828	50	11	39	5	2	5	0
山　东 Shandong	6423	60	46	14	33	1	6	0
广　东 Guangdong	3387	50	11	39	15	0	28	0
广　西 Guangxi	16996	4	4	0	0	0	2	0
海　南 Hainan	25020	21	4	17	0	19	1	0

7-6 全国海洋生态监控区基本情况（2017年）
Basic Condition of the Marine Ecological Monitoring Areas throughout the Country (2017)

生态监控区 Ecological Monitoring Area	所在地 Location	面积（平方千米） Area (km^2)	主要生态系统类型 Major Types of Ecosystem	多样性指数 Diversity Indices		
				浮游植物 Phyto-plankton	大型浮游动物 Macrozoo-plankton	底栖生物 Macrobenthos
双台子河口 Shungtaizi Estuary	辽宁省 Liaoning Province	3000	河口 Estuary			
锦州湾 Jinzhou Bay	辽宁省 Liaoning Province	650	海湾 Bay	1.8	1.9	0.7
滦河口-北戴河 Luanhekou-Beidaihe	河北省 Hebei Province	900	河口 Estuary	2.7	1.9	1.8
渤海湾 Bohai Bay	天津市 Tianjin Municipality	3000	海湾 Bay	1.9	1.3	2.4
莱州湾 Laizhou Bay	山东省 Shandong Province	3770	海湾 Bay	1.9	1.6	2.8
黄河口 Yellow River Estuary	山东省 Shandong Province	2600	河口 Estuary	2.0	0.9	2.3
苏北浅滩 North Jiangsu Bank	江苏省 Jiangsu Province	15400	滩涂湿地 Tidal Flat Wetland	2.5	2.4	1.7
长江口 Yangtze Estuary	上海市 Shanghai Municipality	13668	河口 Estuary	1.6	2.0	2.1

注：生物多样性指数是生物种数和种类间个体数量分配均匀性的综合表现，用Shannon-Wiener多样性指数表征。

Note: Biodiversity index refers to the comprehensive expression of the distributive homogeneity of the number of biological species and the number of individuals between varieties characterized by the Shannon-Wiener biodiversity index.

7-6续表continued

生态监控区 Ecological Monitoring Area	所在地 Location	面积（平方千米）Area (km^2)	主要生态系统类型 Major Types of Ecosystem	多样性指数 Diversity Indices		
				浮游植物 Phyto-plankton	大型浮游动物 Macrozoo-plankton	底栖生物 Macrobenthos
杭州湾 Hangzhou Bay	上海市 浙江省 Shanghai Municipality Zhejiang Province	5000	海湾 Bay	1.8	1.9	0.3
乐清湾 Yueqing Bay	浙江省 Zhejiang Province	464	海湾 Bay	2.3	2.5	1.5
闽东沿岸 Coastal East Fujian	福建省 Fujian Province	5063	海湾 Bay	1.7	2.4	3.0
大亚湾 Daya Bay	广东省 Guangdong Province	1200	海湾 Bay	1.7	2.5	1.9
珠江口 Zhujiang River Mouth	广东省 Guangdong Province	3980	河口 Estuary	1.7	2.3	1.4
广西北海 Beihai, Guangxi	广西壮族自治区 Guangxi Zhuang Nationality Autonomous Region	120	红树林 海草床 Coral Reef Mangroves Seagrass Bed			
北仑河口 Beilun River Mouth	广西壮族自治区 Guangxi Zhuang Nationality Autonomous Region	150	红树林 Mangroves			
海南东海岸 East Coast of Hainan	海南省 Hainan Province	3750	珊瑚礁 海草床 Coral Reef Seagrass Bed			
西沙珊瑚礁 Xisha Coral Reef	海南省 Hainan Province	400	珊瑚礁 Coral Reef			

7-7 沿海地区风暴潮灾害情况（2017年）
Survey of Storm Surges Disasters by Coastal Regions (2017)

受灾地区 Disaster Area	受灾人口 （万人） Disaster-stricken Population (10000 persons)	死亡人数* （人） Death Toll (person)	受灾面积 Disaster-stricken area	
			农田 （千公顷） Disaster-affected farmland (1000 hm^2)	水产养殖 （千公顷） Affected Area of Mariculture (1000 hm^2)
合　计 Total	**171.5**	**6**	**13.8**	**28.1**
山　东 Shandong		0	0.0	0.0
浙　江 Zhejiang		0	0.0	1.1
福　建 Fujian		0	0.0	2.6
广　东 Guangdong	171.5	6	13.8	24.4
广　西 Guangxi		0	0.0	0.0

注：*包括失踪人数。

Notes:*Includes the number of missing people.

7-7 续表 continued

受灾地区 Disaster Area	海岸工程 （千米） Coastal Engineering (km)	房屋 （间） House	船只 （艘） Boat	直接经济损失 （亿元） Direct Economic Loss (100 million yuan)
合 计 Total	**781.8**	**34**	**426**	**55.77**
山 东 Shandong	3.0	0	0	0.06
浙 江 Zhejiang	0.7	0	21	0.87
福 建 Fujian	2.1	0	47	1.21
广 东 Guangdong	776.0	34	358	53.61
广 西 Guangxi	0.0	0	0	0.02

7-8 沿海地区赤潮灾害情况（2017年）
Survey of Red Tide Disasters by Coastal Regions (2017)

时　间 Date	影响区域 Affected Area	最大面积 （平方千米） Max Area (km^2)
合　计 **Total**		**2 488**
2月27日至3月17日 Feb.27-Mar.17	茂名水东湾附近海域 Sea area near Shuidong Bay of Maoming	495
3月14日至31日 Mar.14-Mar.31	湛江海湾大桥以南至金沙滩附近海域 Sea area south of Zhanjiang Bay Bridge to Golden Beach	175
3月21日至4月12日 Mar.21-Apr.12	天津中心渔港附近海域 Sea area near Tianjin central fishing port	160
3月23日至4月6日 Mar.23-Apr.6	湛江雷州半岛水尾以南至角尾对出海域 Sea area south of Zhanjiang's Leizhou Peninsula to Jiaowei	118
3月23日至4月6日 Mar.23-Apr.6	湛江东海岛通明出海口以东至东南码头附近海域 Sea area east of Tongming Estuary of Zhanjiang's Donghai Island to the southeast pier	100
5月17日至19日 May.17-May.19	连云港排淡河口至埒子河口邻近海域 Sea area near Lianyungang's Paidan River Estuary to Liezi River Estuary	100

注：本表仅列出最大面积超过100平方千米（含）的赤潮过程。

Note: This table only lists the red tide processes with the maximum area each exceeding 100 km^2.

7-8 续表 continued

时 间 Date	影响区域 Affected Area	最大面积（平方千米） Max Area (km^2)
5月20日至24日 May.20-May.24	渔山海域 Yushan sea area	220
6月16日至7月6日 Jun.16-Jul.6	渔山列岛与檀头山之间中部海域 Central sea area between Yushan Islands and Tantou Mountain	420
6月20日至26日 Jun.20-Jun.26	舟山朱家尖至虾峙岛以东海域 Eastern sea area from Zhoushan's Zhujiajian to Xiazhi Island	180
6月20日至30日 Jun.20-Jun.30	南韭山西部海域 Western sea area of Nanjiu Mountain	120
6月27日至30日 Jun.27-Jun.30	洛屿岛以外海域及玉环岛披山海域 Sea area off Luoyu Island and sea area near Yuhuan Pishan Island	300
7月7日至11日 Jul.7-Jul.11	舟山朱家尖以东海域 Sea area east of Zhujiajian, Zhoushan	100

主要统计指标解释

1. 工业废水排放量　指经过企业厂区所有排放口排到企业外部的工业废水量。包括生产废水、外排的直接冷却水、超标排放的矿井地下水和与工业废水混排的厂区生活污水,不包括外排的间接冷却水(清污不分流的间接冷却水应计算在内)。

2. 直接排入海的工业废水量　指经企业位于海边的排放口，直接排入海中的废水量。直接排入是指废水经过工厂的排污口直接排入海，而未经过城市下水道或其他中间体，也不受其他水体的影响。

3. 工业废水处理量　指报告期内各种水治理设施实际处理的工业废水量,包括处理后外排的和处理后回用的工业废水量，虽经处理但未达到国家或地方排放标准的废水量也应计算在内。计算时，如遇车间和厂排放口均有治理设施，并对同一废水分级处理时，不应重复计算工业废水处理量。

4. 当年施工项目数　指报告期内由国家、部门、地方或企业单位安排开工的，并以治理废水、废气、固体废物、噪声和其他（如电磁波、恶臭等）环境污染为目的的环境治理工程的总数。不包括“三同时”项目。

5. 当年竣工项目数　指报告期内竣工投入运行的治理废水、废气、固体废物、噪声及其他污染的环境工程项目的总数。

Explanatory Notes on Main Statistical Indicators

1. Volume of Industrial Waste Water Discharged　refers to the quantity of industrial waste water discharged externally through all the outlets in the factory area of the enterprise, including the waste water from production, externally discharged direct cooling water, mine-shaft groundwater discharged exceeding the set standard and the domestic sewage of the factory area discharged together with the industrial waste water, but not including the indirect cooling water discharge externally.

2. Volume of Industrial Waste Water Discharged Directly to the Sea　refers to the quantity of waste water directly discharged into the sea through the outlets of the enterprise by the sea. Direct discharge means the direct discharge into the sea of waste water through the outlets of the factory, which is not discharged via the urban sewers or other intermediates and is not affected by other water bodies.

3. Volume of Industrial Waste Water Treated　refers to the industrial waste water volume actually treated by various water treatment facilities in the period covered by the report, including the quantity of the industrial waste water discharged and reused after treatment. The amount of the waste water which is not up to the state or local standard of discharge upon treatment should be included. If the workshops and outlets of the factory are provided with treatment facilities and carry out graded treatment of the same waste water, the processed volume of industrial waste water cannot be calculated repeatedly.

4. Number of Construction Projects in the Current Year refers to the total number of environmental pollution control projects for the purpose of controlling waste water, waste gas, solid wastes, noise and other environmental pollutions (such as electromagnetic wave, offensive odor) started by the state, governmental departments, local governments or enterprises in the period covered by the report. It does not include the projects of "three simultaneousness".

5. Number of Pollution Treatment Projects Completed in the Current Year refers to the total number of environmental engineering projects for controlling waste water, waste gas, solid wastes, noise and other pollutions which are completed and put into operation in the period covered by the report.

8

海洋行政管理及公益服务

Marine Administration and Public-Good Service

8-1 海域使用管理情况（2017年）
Sea Area Use Management (2017)

地 区 Region	新增宗海数量 （宗） New Sea Parcels (parcel)	新增宗海面积 （公顷） Area of New Sea Parcel (hm^2)	海域使用金征收金额 （万元） Charge for Sea Area Use (10000 yuan)
全国总计 National Total	**1831**	**168111.6**	**648519.4**
天 津 Tianjin	8	895.6	38667.8
河 北 Hebei	414	14868.9	18774.0
辽 宁 Liaoning	437	94351.7	65425.1
上 海 Shanghai	9	419.5	33748.4
江 苏 Jiangsu	131	22291.7	45416.7
浙 江 Zhejiang	259	3772.7	102701.2
福 建 Fujian	89	3710.1	64649.2
山 东 Shandong	220	22388.9	73439.0
广 东 Guangdong	197	4098.3	156151.3
广 西 Guangxi	42	751.5	8655.4
海 南 Hainan	25	562.7	35812.6
其 他 Others	0	0.0	5078.8

注：其他为沿海省（自治区、直辖市）管理海域以外（指渤海中部海域）。

Note: Others are the sea areas outside the control of coastal provinces, autonomous regions and municipalities directly under the Central Government(Referring to the mid-Bohai Sea area).

8-2 依法批准无居民海岛开发利用情况
Development and Utilization of Uninhabited Islands Legally Approved

地 区 Region	海岛名称 Island Name	用 途 Use	用岛面积（公顷） Island Area（hm^2）	批准年份 Approved Year
浙 江 Zhejiang	北一江山岛 North Yijiangshan Island	公益服务 Public-Good Service	0.5	2017
	担峙岛 Danzhi Island	交通运输 Communications and Transport	0.4	2017
福 建 Fujian	黄官岛 Huangguan Island	旅游娱乐 Tourism and Recreation	0.3	2016
	西洛岛 Xiluo Island	道路广场 Road and Square	0.6	2017
	火烧屿 Huoshaoyu	施工期用岛 Construction	3.1	2017
	大兔屿 Datuyu	施工期用岛 Construction	5.1	2017
	横山墩 Hengshandun	桥梁 Bridge	0.2	2017
广 西 Guangxi	蟾蜍墩 Chanchudun	桥梁 Bridge	0.3	2017
	草埠岛 Caobu Island	桥梁 Bridge	0.4	2017

注：数据来源于《2017年海岛统计调查公报》。

Note: The data come from the *Bulletin of Sea Islands Statistical Survey 2017.*

8-3 沿海地区海滨观测台站分布概况（2017年）
Distribution of Coastal Observation Stations by Coastal Regions (2017)

单位：个 (unit)

地 区 Region	合 计 Total	海洋站 Marine Station	验潮站[①] Tide Station	气象台站 Meteorological Station	地震台站 Seismic Station
合 计 Total	**1117**	**156**	**252**	**556**	**152**
天 津 Tianjin	**27**	3		15	9
河 北 Hebei	**46**	7		3	36
辽 宁 Liaoning	**149**	13	3	115	18
上 海 Shanghai	**131**	9	64	56	2
江 苏 Jiangsu	**89**	13	26	34	15
浙 江 Zhejiang	**113**	25	37	45	6
福 建 Fujian	**178**	17	11	136	14
山 东 Shandong	**125**	26	21	52	26
广 东 Guangdong	**148**	21	82	33	12
广 西 Guangxi	**40**	7	5	19	9
海 南 Hainan	**71**	15	3	48	5

注：①潮流量观测站43处，潮水位观测站209处。

Notes: ①There are 43 tidal current observation stations and 209 tidal level observation stations.

8-4 海洋预报服务概况（国家级，2017年）
Marine Forecast Service (National Level, 2017)

单位：次 (time)

预报项目 Item	数值预报 Numerical Forecast			
	预报服务次数 Frequency	发布次数 Frequency of Release		
		广播电视 Radio and TV	互联网 Internet	纸 质 Paper Media
合　计 Total	**7663**	**365**	**7650**	**13**
海　浪 Sea Wave	730	365	730	
海　温 Sea Surface Temperature	2190		2190	
潮　汐 Tide				
海　流 Sea Current	2190		2190	
海平面 Sea Level				
盐　度 Salinity	2190		2190	
赤　潮 Red Tide				
滨海旅游 Coastal Tourism				
海　冰 Sea Ice	261		248	13
绿　潮 Green Seaweed	90		90	
溢　油 Oil Spill				
厄尔尼诺 El Niño				
专　项 Special Item	8		8	
其　他 Others	4		4	

8-4 续表 continued

预报项目 Item	统计预报 Statistical Forecast			
	预报服务次数 Frequency	发布次数 Frequency of Release		
		广播电视 Radio and TV	互联网 Internet	纸 质 Paper Media
合 计 Total	**71572**	**10565**	**32195**	**38002**
海 浪 Sea Wave	964	730	964	234
海 温 Sea Surface Temperature	2610	2190	400	
潮 汐 Tide	18120	5940		18000
海 流 Sea Current	18500	780		18500
海平面 Sea Level	500			500
盐 度 Salinity				
赤 潮 Red Tide	18		18	
滨海旅游 Coastal Tourism	2197	916	2197	730
海 冰 Sea Ice	113	9	78	26
绿 潮 Green Seaweed				
溢 油 Oil Spill				
厄尔尼诺 EL Niño	12			12
专 项 Special Item	28538		28538	
其 他 Others				

8-5 海洋观测调查情况（2017年）
Basic Statistics on Ocean Observation (2017)

项目 Item	合 计 **Total**	志愿船观测 Volunteer Ship Observation	断面观测 Sectional Observation	台站观测 Station Observation	浮标观测 Buoy Monitoring
站点数（个） Number of Stations (unit)	**354**	59	119	124	52
观测数据（MB） Data (MB)	**16325.4**	2956.4	364.8	12707.8	296.4

8-6 海洋调查概况（2017年）
Marine Survey Statistics (2017)

调查类别 Name	站点数（个） Number of Stations (unit)	船舶数（艘） Number of Ships (unit)	项目数（个） Number of Items (unit)	实际获得数据（个） Quantity of Data Actually Obtained (unit)	发布通（公、简）报量（期） Quantity of Circulars (Bulletin, Brief Reports) (unit)
合 计 **Total**	**2171**	**134**	**423**		**73**
大洋调查 Oceanic Survey	717	13	38		
极地调查 Polar Survey	41	5	131	165498	
专项调查 Special Survey	941	10	39	24101	65
其他调查 Other Surveys	472	106	215	3312682	8

注：数据来源于自然资源部北海局、自然资源部东海局、自然资源部南海局、中国大洋矿产资源研究开发协会办公室、中国极地研究中心、第一海洋研究所、第三海洋研究所。

Note: The data come from the North Sea, East China Sea and South China Sea Bureaus of the Ministry of Natural Resources, Office of the China Ocean Mineral Resources Research and Development Association, China Polar Research Centre, First Institute of Oceanography and Third Institute of Oceanography.

8-7 海洋标准化监督管理情况（2017年）
Supervision and Management of Marine Standardization (2017)

单位：项 (item)

指　标 Item	指 标 值 Data
标准制（修）定 Standards Formulation(Revision)	
国家标准 National Standards	48
行业标准 Professional Standards	306
标准化监督管理 Standardized Supervision and Management	
标准立项 Standards Proposal Approval	
国家标准 National Standards	41
行业标准 Professional Standards	157
标准审查 Standards Examination	
国家标准 National Standards	48
行业标准 Professional Standards	306
标准发布 Standards Issuing	
国家标准 National Standards	37
行业标准 Professional Standards	17
标准出版 Standards Publication	
国家标准 National Standards	37
行业标准 Professional Standards	17
标准实施监督检查（次） Supervision and Examination of Standards Implementation (time)	1

主要统计指标解释

1. 断面观测　每年定期利用船舶在沿海设定的断面上进行海洋水文、气象、生物、化学等项目的监测活动。

2. 浮标观测　在海上固定站位获取长期、连续海洋环境观测资料的海上锚定资料浮标。

3. 大洋调查　以大洋科考、研究为目的的远洋调查。

4. 专项调查　为完成国家专项任务进行的海洋调查。

5. 国家标准　针对海洋领域内需要在全国范围内统一的有关技术要求所制定的国家标准。海洋国家标准由国家标准化主管部门统一批准、编号和发布。

6. 行业标准　对没有海洋国家标准而又需要在海洋领域内统一的技术要求所制定的标准。海洋行业标准由国家海洋局统一批准、编号和发布。

Explanatory Notes on Main Statistical Indicators

1. Sectional Monitoring refers to the monitoring activities concerning such items as marine hydrology, meteorology, biology and chemistry carried out regularly every year on the sections set in the coastal area.

2. Buoy Monitoring refers to the monitoring carried out by the offshore mooring data buoys which acquire long-term, continuous marine environmental observations at the fixed stations at sea.

3. Oceanic Survey refers to the oceanic surveys aimed at the oceanic scientific investigations and research.

4. Special Survey refers to the oceanic investigation for the purpose of fulfilling the state′s special tasks.

5. National Standards refers to the standards formulated in view of the relevant technical requirements in the marine field that need to be unified throughout the country. The marine national standards are approved, numbered and issued uniformly by the state department responsible for standardization.

6. Professional Standards refers to the standards formulated for the technical requirements which have no national standards but need to be unified in the marine field. The marine professional standards are approved, numbered and issued by the State Oceanic Administration.

9

全国及沿海社会经济

National and Coastal Socioeconomy

9-1 国内生产总值
Gross Domestic Product

单位：亿元 (100 million yuan)

年 份 Year	国内生产总值 Gross Domestic Product	第一产业 Primary Industry	第二产业 Secondary Industry	第三产业 Tertiary Industry
2001	110863.1	15502.5	49660.7	45700.0
2002	121717.4	16190.2	54105.5	51421.7
2003	137422.0	16970.2	62697.4	57754.4
2004	161840.2	20904.3	74286.9	66648.9
2005	187318.9	21806.7	88084.4	77427.8
2006	219438.5	23317.0	104361.8	91759.7
2007	270232.3	27788.0	126633.6	115810.7
2008	319515.5	32753.2	149956.6	136805.8
2009	349081.4	34161.8	160171.7	154747.9
2010	413030.3	39362.6	191629.8	182038.0
2011	489300.6	46163.1	227038.8	216098.6
2012	540367.4	50902.3	244643.3	244821.9
2013	595244.4	55329.1	261956.1	277959.3
2014	643974.0	58343.5	277571.8	308058.6
2015	689052.1	60862.1	282040.3	346149.7
2016	743585.5	63672.8	296547.7	383365.0
2017	827121.7	65467.6	334622.6	427031.5

9-2 国内生产总值增长速度
Growth Rate of Gross Domestic Product

单位：% (%)

年 份 Year	国内生产总值 Gross Domestic Product	第一产业 Primary Industry	第二产业 Secondary Industry	第三产业 Tertiary Industry
2001	8.3	2.6	8.5	10.3
2002	9.1	2.7	9.9	10.5
2003	10.0	2.4	12.7	9.5
2004	10.1	6.1	11.1	10.1
2005	11.4	5.1	12.1	12.4
2006	12.7	4.8	13.5	14.1
2007	14.2	3.5	15.1	16.1
2008	9.7	5.2	9.8	10.5
2009	9.4	4.0	10.3	9.6
2010	10.6	4.3	12.7	9.7
2011	9.5	4.2	10.7	9.5
2012	7.9	4.5	8.4	8.0
2013	7.8	3.8	8.0	8.3
2014	7.3	4.1	7.4	7.8
2015	6.9	3.9	6.2	8.2
2016	6.7	3.3	6.3	7.7
2017	6.9	3.9	6.1	8.0

注：本表按可比价格计算（上年为基期）。

Note:This table is calculated at the comparable price (with the previous year as the base period).

9-3 沿海地区生产总值（2017年）
Gross Regional Product of Coastal Regions (2017)

单位：亿元　　(100 million yuan)

地 区 Region	地区生产总值 Gross Regional Product	第一产业 Primary Industry	第二产业 Secondary Industry	第三产业 Tertiary Industry
合 计 Total	**461753.0**	**25791.5**	**197609.8**	**238351.9**
天 津 Tianjin	18549.2	169.0	7593.6	10786.6
河 北 Hebei	34016.3	3130.0	15846.2	15040.1
辽 宁 Liaoning	23409.2	1902.3	9199.8	12307.2
上 海 Shanghai	30633.0	110.8	9330.7	21191.5
江 苏 Jiangsu	85869.8	4045.2	38654.9	43169.7
浙 江 Zhejiang	51768.3	1933.9	22232.1	27602.3
福 建 Fujian	32182.1	2215.1	15354.3	14612.7
山 东 Shandong	72634.1	4832.7	32942.8	34858.6
广 东 Guangdong	89705.2	3611.4	38008.1	48085.7
广 西 Guangxi	18523.3	2878.3	7450.9	8194.1
海 南 Hainan	4462.5	962.8	996.4	2503.4

9-4 沿海地区生产总值增长速度
Growth Rate of Gross Regional Product of Coastal Regions

单位：% (%)

地 区 Region	2010	2011	2012	2013	2014	2015	2016	2017
天 津 Tianjin	17.4	16.4	13.8	12.5	10.0	9.3	9.1	3.6
河 北 Hebei	12.2	11.3	9.6	8.2	6.5	6.8	6.8	6.6
辽 宁 Liaoning	14.2	12.2	9.5	8.7	5.8	3.0	-2.5	4.2
上 海 Shanghai	10.3	8.2	7.5	7.7	7.0	6.9	6.9	6.9
江 苏 Jiangsu	12.7	11.0	10.1	9.6	8.7	8.5	7.8	7.2
浙 江 Zhejiang	11.9	9.0	8.0	8.2	7.6	8.0	7.6	7.8
福 建 Fujian	13.9	12.3	11.4	11.0	9.9	9.0	8.4	8.1
山 东 Shandong	12.3	10.9	9.8	9.6	8.7	8.0	7.6	7.4
广 东 Guangdong	12.4	10.0	8.2	8.5	7.8	8.0	7.5	7.5
广 西 Guangxi	14.2	12.3	11.3	10.2	8.5	8.1	7.3	7.1
海 南 Hainan	16.0	12.0	9.1	9.9	8.5	7.8	7.5	7.0

注：本表按可比价格计算（上年为基期）。

Note: This table is calculated at the comparable price (with the previous year as the base period).

9-5 沿海城市生产总值（2016年）
Gross Regional Product of Coastal Cities, 2016

单位：亿元 (100 million yuan)

沿海城市 Coastal City		地区生产总值 Gross Regional Product	第一产业 Primary Industry	第二产业 Secondary Industry	第三产业 Tertiary Industry
合　计	**Total**	**251993.7**	**11830.6**	**106726.5**	**133436.6**
天　津	**Tianjin**	**17885.4**	**220.2**	**7571.4**	**10093.8**
河　北	**Hebei**	**11249.0**	**1103.6**	**5726.3**	**4419.1**
唐　山	Tangshan	6354.9	599.0	3499.9	2256.0
秦皇岛	Qinhuangdao	1349.4	196.0	468.6	684.8
沧　州	Cangzhou	3544.7	308.6	1757.8	1478.3
辽　宁	**Liaoning**	**11410.6**	**1137.7**	**4580.1**	**5692.8**
大　连	Dalian	6810.3	462.8	2849.9	3497.6
丹　东	Dandong	751.2	127.8	231.1	392.3
锦　州	Jinzhou	1032.9	205.6	353.3	474.0
营　口	Yingkou	1156.3	110.9	471.5	573.9
盘　锦	Panjin	1012.6	121.3	449.3	442.0
葫芦岛	Huludao	647.3	109.3	225.0	313.0
上　海	**Shanghai**	**28178.7**	**109.5**	**8406.3**	**19662.9**
江　苏	**Jiangsu**	**13720.8**	**1202.2**	**6270.2**	**6248.4**
南　通	Nantong	6768.2	366.7	3170.3	3231.2
连云港	Lianyungang	2376.5	301.6	1049.9	1025.0
盐　城	Yancheng	4576.1	533.9	2050.0	1992.2
浙　江	**Zhejiang**	**38892.9**	**1471.3**	**17287.3**	**20134.3**
杭　州	Hangzhou	11313.7	304.2	4120.9	6888.6
宁　波	Ningbo	8686.5	302.1	4455.3	3929.1
温　州	Wenzhou	5101.7	139.6	2096.5	2865.6
嘉　兴	Jiaxing	3862.1	136.9	2010.5	1714.7
绍　兴	Shaoxing	4789.1	207.7	2398.3	2183.1
舟　山	Zhoushan	1241.2	126.7	510.0	604.5
台　州	Taizhou	3898.6	254.1	1695.8	1948.7
福　建	**Fujian**	**23200.6**	**1542.1**	**11312.0**	**10346.5**
福　州	Fuzhou	6197.7	492.3	2590.4	3115.0
厦　门	Xiamen	3784.3	23.2	1544.6	2216.5
莆　田	Putian	1823.5	126.5	1022.5	674.5
泉　州	Quanzhou	6646.7	198.5	3886.7	2561.5
漳　州	Zhangzhou	3125.3	415.6	1461.1	1248.6
宁　德	Ningde	1623.1	286.0	806.7	530.4

注：本表各省数据为沿海城市合计数。

Note: The data for the provinces are the total of coastal cities.

9-5 续表 continued

沿海城市 Coastal City		地区生产总值 Gross Regional Product	第一产业 Primary Industry	第二产业 Secondary Industry	第三产业 Tertiary Industry
山　东	**Shandong**	**33424.1**	**2044.2**	**15803.4**	**15576.5**
青　岛	Qingdao	10011.3	371.0	4160.7	5479.6
东　营	Dongying	3479.6	121.9	2163.1	1194.6
烟　台	Yantai	6925.7	467.5	3461.7	2996.5
潍　坊	Weifang	5522.7	475.3	2559.8	2487.6
威　海	Weihai	3212.2	229.3	1463.4	1519.5
日　照	Rizhao	1802.5	147.0	851.9	803.6
滨　州	Binzhou	2470.1	232.2	1142.8	1095.1
广　东	**Guangdong**	**69513.4**	**2391.6**	**28057.1**	**39064.7**
广　州	Guangzhou	19547.5	239.3	5751.6	13556.6
深　圳	Shenzhen	19492.7	7.2	7780.5	11705.0
珠　海	Zhuhai	2226.4	43.5	1079.9	1103.0
汕　头	Shantou	2080.9	107.2	1051.0	922.7
江　门	Jiangmen	2418.9	189.0	1150.8	1079.1
湛　江	Zhanjiang	2584.5	497.6	985.9	1101.0
茂　名	Maoming	2636.7	435.9	1058.7	1142.1
惠　州	Huizhou	3412.2	171.7	1837.5	1403.0
汕　尾	Shanwei	828.6	129.8	368.7	330.1
阳　江	Yangjiang	1270.7	219.0	520.8	530.9
东　莞	Dongguan	6827.7	24.2	3173.2	3630.3
中　山	Zhongshan	3202.9	68.3	1677.3	1457.3
潮　州	Chaozhou	976.8	70.4	502.8	403.6
揭　阳	Jieyang	2006.9	188.5	1118.4	700.0
广　西	**Guangxi**	**2784.9**	**477.5**	**1384.3**	**923.1**
北　海	Beihai	1006.7	174.8	516.1	315.8
防城港	Fangchenggang	676.1	82.6	386.3	207.2
钦　州	Qinzhou	1102.1	220.1	481.9	400.1
海　南	**Hainan**	**1733.3**	**130.7**	**328.1**	**1274.5**
海　口	Haikou	1257.7	63.9	233.6	960.2
三　亚	Sanya	475.6	66.8	94.5	314.3

9-6 县级单位主要统计指标
Main Indicators of Regions at County Level

单位：万元 (10000 yuan)

沿海县 Coastal County		第一产业增加值 Value-added of Primary Industry		第二产业增加值 Value-added of Secondary Industry	
		2015	2016	2015	2016
合　计	**Total**	**58361610**	**62038773**	**244177321**	**252628531**
河　北	**Hebei**	**3476907**	**3667890**	**7872780**	**8018803**
丰　南	Fengnan	471462	483438	3714337	3595292
滦　南	Luannan	856425	920000	944010	1029894
乐　亭	Laoting	863597	939208	985083	1069062
昌　黎	Changli	556470	589385	734255	776737
抚　宁	Funing	333399	343787	280367	261503
黄　骅	Huanghua	324994	312300	1028216	1065530
海　兴	Haixing	70560	79772	186512	220785
辽　宁	**Liaoning**	**7193167**	**7220980**	**19781223**	**13615697**
长　海	Changhai	470014	501088	72123	57998
瓦房店	Wafangdian	943210	952812	5909195	4744099
普兰店	Pulandian	980254	853258	3921754	2131645
庄　河	Zhuanghe	1270383	1311650	3078274	2694275
东　港	Donggang	775879	677934	1325368	687731
凌　海	Linghai	552864	565263	1331563	489018
盖　州	Gaizhou	377750	448475	775337	486921
大　洼	Dawa	676612	665068	1793923	1232349
盘　山	Panshan	498065	513869	676643	539685
绥　中	Suizhong	449281	485290	465476	400651
兴　城	Xingcheng	198855	246273	431567	151325

注：本表各省数据为沿海县合计数。

Note: The data for the provinces are the total of coastal counties.

9-6 续表1 continued

沿海县 Coastal County		第一产业增加值 Value-added of Primary Industry		第二产业增加值 Value-added of Secondary Industry	
		2015	2016	2015	2016
江　苏	**Jiangsu**	**8138946**	**8527100**	**30770161**	**32972000**
海　安	Hai'an	537710	559600	3231845	3541500
如　东	Rudong	649665	678700	3148317	3405700
启　东	Qidong	654460	665800	3891337	4228500
海　门	Haimen	517975	532800	4715945	5045300
赣　榆	Ganyu	699700	765500	2302700	2455600
东　海	Donghai	618900	672800	1737100	1865700
灌　云	Guanyun	598100	641000	1335200	1431200
灌　南	Guannan	482300	515100	1368600	1447800
响　水	Xiangshui	403500	414200	1136752	1267700
滨　海	Binhai	559741	579900	1489758	1567200
射　阳	Sheyang	784100	810100	1476102	1557800
东　台	Dongtai	879195	914700	2792105	2920300
大　丰	Dafeng	753600	776900	2144400	2237700
浙　江	**Zhejiang**	**6606480**	**7126948**	**56675069**	**62310168**
象　山	Xiangshan	613666	669957	1846924	1971666
宁　海	Ninghai	411164	438956	2242975	2512165
余　姚	Yuyao	423578	449230	4683538	5075861
慈　溪	Cixi	497096	526762	6560527	7622696
奉　化	Fenghua	290288	300173	1432944	3008866
洞　头	Dongtou	54000	60400	290000	296500
平　阳	Pingyang	144839	159734	1458333	1531673
苍　南	Cangnan	296232	319279	1754813	1827118
瑞　安	Rui'an	206440	222692	3290000	3387531
乐　清	Yueqing	209106	214047	3909661	4173567
海　盐	Haiyan	207391	206315	2288589	2460805
海　宁	Haining	215675	216780	3844508	4147424
平　湖	Pinghu	147090	151937	2846236	3056817

9-6 续表2 continued

沿海县 Coastal County	第一产业增加值 Value-added of Primary Industry		第二产业增加值 Value-added of Secondary Industry	
	2015	2016	2015	2016
越 城 Yuecheng				
柯 桥 Keqiao	339097	357444	6446215	6596925
上 虞 Shangyu	426260	453586	3909580	4230236
岱 山 Daishan	324500	368783	1065014	1165474
嵊 泗 Shengsi	232345	268900	130184	136900
玉 环 Yuhuan	293853	327746	2407779	2479653
三 门 Sanmen	247308	274679	642074	671502
温 岭 Wenling	623656	695808	3563446	3664388
临 海 Linhai	402896	443740	2061729	2292401
福 建 Fujian	**8034107**	**9063854**	**44880248**	**50966786**
连 江 Lianjiang	1199308	1389765	1391522	1490113
罗 源 Luoyuan	336836	388054	1120198	1175629
平 潭 Pingtan	359540	358178	594296	598470
福 清 Fuqing	909498	1027584	192445	4211017
长 乐 Changle	432828	513711	3731900	3953424
仙 游 Xianyou	305583	323681	1590065	1693610
惠 安 Hui'an	289695	302919	5025210	5489775
金 门 Jinmen				
石 狮 Shishi	200167	235686	3835133	3705400
晋 江 Jinjiang	195941	213221	10324385	10600289
南 安 Nan'an	247201	281660	5180200	5412778
云 霄 Yunxiao	289943	301268	744143	832903
漳 浦 Zhangpu	622951	744264	1291365	1200892
诏 安 Zhao'an	392790	444579	835472	939360
东 山 Dongshan	308588	336618	742825	823677
龙 海 Longhai	575369	644173	3636115	3938439
霞 浦 Xiapu	516629	596180	563927	596710
福 安 Fu'an	438556	491584	2214758	2314249
福 鼎 Fuding	412684	470729	1866289	1990051

9-6 续表3 continued

沿海县 Coastal County		第一产业增加值 Value-added of Primary Industry		第二产业增加值 Value-added of Secondary Industry	
		2015	2016	2015	2016
山 东	**Shandong**	**9538043**	**9412365**	**58984260**	**57937907**
胶 州	Jiaozhou	532000	499432	5178000	5381500
即 墨	Jimo	601922	622230	5980400	6353700
胶 南	Jiaonan				
垦 利	Kenli	714657	202355	5132557	2343369
利 津	Lijin	270753	280708	1326837	1342555
广 饶	Guangrao	421601	428831	5133361	5148259
长 岛	Changdao	361040	377472	36817	38845
龙 口	Longkou	361854	383940	6064091	6336412
莱 阳	Laiyang	448185	477576	1522801	1609225
莱 州	Laizhou	681153	711903	3734380	3901537
蓬 莱	Penglai	280713	291979	2485597	2599396
招 远	Zhaoyuan	400114	423432	3383443	3538588
海 阳	Haiyang	629594	665969	1045117	1102984
寿 光	Shouguang	931600	973300	3675400	3641300
昌 邑	Changyi	347350	369533	2040400	2025500
文 登	Wendeng	580521	613394	3329730	3457704
荣 成	Rongcheng	831341	874640	4747443	4809788
乳 山	Rushan	395467	418643	2266586	2356500
无 棣	Wudi	361158	386810	1283785	1318457
沾 化	Zhanhua	387020	410218	617515	632288
广 东	**Guangdong**	**9181658**	**10177065**	**20662373**	**22116029**
南 澳	Nan’ao	39438	43199	55816	54292
台 山	Taishan	564884	624941	1725306	1859081
恩 平	Enping	200362	200511	515142	534754
遂 溪	Suixi	991268	1058535	772990	788193
徐 闻	Xuwen	677456	770047	139451	109195

9-6 续表4 continued

沿海县 Coastal County		第一产业增加值 Value-added of Primary Industry		第二产业增加值 Value-added of Secondary Industry	
		2015	2016	2015	2016
廉　江	Lianjiang	947462	1048555	1775284	2080688
雷　州	Leizhou	950578	1063153	383995	326588
吴　川	Wuchuan	280913	297939	953563	1079308
电　白	Dianbai	1052795	1199070	2200606	2343229
惠　东	Huidong	426805	499145	2464554	2861671
海　丰	Haifeng	358032	367697	1217964	1278739
陆　丰	Lufeng	491261	553234	1048885	1054848
阳　西	Yangxi	535615	584053	748148	773851
阳　东	Yangdong	436987	488108	1586331	1629851
饶　平	Raoping	397902	450060	980189	1024019
揭　东	Jiedong	321835	360828	2695076	2866617
惠　来	Huilai	508065	567990	1399073	1451105
广　西	**Guangxi**	**943151**	**1035446**	**887990**	**984526**
合　浦	Hepu	786996	865094	525725	600443
东　兴	Dongxing	156155	170352	362265	384083
海　南	**Hainan**	**5249151**	**5807125**	**3663217**	**3706615**
琼　海	Qionghai	721525	789286	273935	292845
儋　州	Danzhou				
文　昌	Wenchang	669711	745728	413954	435620
万　宁	Wanning	521005	599500	354400	375500
东　方	Dongfang	379493	429642	687921	605527
澄　迈	Chengmai	649517	731382	1092000	1065786
临　高	Lingao	1016172	1078792	91841	99864
昌　江	Changjiang	250098	283751	376988	420552
乐　东	Ledong	632074	709399	128746	151792
陵　水	Lingshui	409556	439645	243432	259129

9-6 续表5 continued

沿海县 Coastal County		公共财政收入 Public Revenue of Local Governments		公共财政支出 Public Expenditure of Local Governments	
		2015	2016	2015	2016
合　计	**Total**	**38107159**	**38823648**	**60155149**	**63094682**
河　北	**Hebei**	**831030**	**859499**	**1887392**	**1996759**
丰　南	Fengnan	290948	315208	410245	458451
滦　南	Luannan	95589	102080	254088	274530
乐　亭	Laoting	111160	122301	288112	292601
昌　黎	Changli	91359	100018	249471	277683
抚　宁	Funing	81506	30683	203111	163824
黄　骅	Huanghua	130088	156666	344852	366513
海　兴	Haixing	30380	32543	137513	163157
辽　宁	**Liaoning**	**1742812**	**1718329**	**3999912**	**4351543**
长　海	Changhai	43609	46706	89664	137499
瓦房店	Wafangdian	376450	434051	621085	759473
普兰店	Pulandian	224795	140060	379886	371189
庄　河	Zhuanghe	229103	234508	449470	565032
东　港	Donggang	132195	117700	471413	391899
凌　海	Linghai	85678	77675	308553	288533
盖　州	Gaizhou	101239	88693	306272	340073
大　洼	Dawa	263060	290706	456052	532111
盘　山	Panshan	76486	99829	286858	289916
绥　中	Suizhong	90532	96000	312446	375289
兴　城	Xingcheng	119665	92401	318213	300529

9-6 续表6 continued

沿海县 Coastal County			公共财政收入 Public Revenue of Local Governments		公共财政支出 Public Expenditure of Local Governments	
			2015	2016	2015	2016
江	苏	**Jiangsu**	**6721871**	**5517700**	**9931391**	**9448648**
海	安	Hai'an	620566	575767	843989	813491
如	东	Rudong	585447	544084	938165	1021563
启	东	Qidong	768609	710317	862358	890262
海	门	Haimen	783980	724112	855534	857533
赣	榆	Ganyu	445078	250181	755798	602247
东	海	Donghai	410163	226068	687167	580905
灌	云	Guanyun	393199	215179	618031	496431
灌	南	Guannan	388511	224336	609084	480435
响	水	Xiangshui	325412	296024	508150	501274
滨	海	Binhai	386540	340700	694488	708929
射	阳	Sheyang	205588	215800	593838	645980
东	台	Dongtai	715512	602566	1017151	955830
大	丰	Dafeng	693266	592566	947638	893768
浙	江	**Zhejiang**	**9775208**	**10498276**	**12582934**	**14278131**
象	山	Xiangshan	379485	380730	613051	613659
宁	海	Ninghai	415712	487899	602441	677589
余	姚	Yuyao	750989	811633	925207	936884
慈	溪	Cixi	1122589	1320963	1254760	1457152
奉	化	Fenghua	338192	370515	607125	630695
洞	头	Dongtou	54172	65182	187916	203418
平	阳	Pingyang	256688	277720	457164	546966
苍	南	Cangnan	320667	318176	544021	701325
瑞	安	Rui'an	547214	590481	733731	927977
乐	清	Yueqing	955544	722325	702225	862087
海	盐	Haiyan	316288	351779	400298	435080
海	宁	Haining	691221	720018	765612	780895
平	湖	Pinghu	505780	567891	538324	564144

9-6 续表7 continued

沿海县 Coastal County	公共财政收入 Public Revenue of Local Governments		公共财政支出 Public Expenditure of Local Governments	
	2015	2016	2015	2016
越　城 Yuecheng				
柯　桥 Keqiao	978488	1060220	985258	989490
上　虞 Shangyu	535837	596523	600132	650296
岱　山 Daishan	126037	140863	355707	381482
嵊　泗 Shengsi	69010	67301	210244	235991
玉　环 Yuhuan	344987	426217	481184	543880
三　门 Sanmen	144987	155766	305551	329046
温　岭 Wenling	541213	616886	726079	943353
临　海 Linhai	380108	449188	586904	866722
福　建 Fujian	**6582710**	**7666398**	**9224606**	**9955999**
连　江 Lianjiang	381306	430703	460637	568426
罗　源 Luoyuan	108899	88763	201685	254358
平　潭 Pingtan	250156	535172	1241429	1080495
福　清 Fuqing	812583	886796	698100	786825
长　乐 Changle	336452	553697	473385	508420
仙　游 Xianyou	281689	278174	427466	473827
惠　安 Hui'an	330203	419220	545591	620485
金　门 Jinmen				
石　狮 Shishi	592933	600066	483711	486009
晋　江 Jinjiang	1172008	1206806	1297465	1374144
南　安 Nan'an	627698	662889	621394	677898
云　霄 Yunxiao	74722	101238	253458	252406
漳　浦 Zhangpu	315538	267586	480195	507436
诏　安 Zhao'an	80294	81922	271449	275299
东　山 Dongshan	182712	173970	277020	271257
龙　海 Longhai	480409	817990	442963	714038
霞　浦 Xiapu	116707	120084	274111	311210
福　安 Fu'an	235026	186314	418564	406995
福　鼎 Fuding	203375	255008	355983	386471

9-6 续表8 continued

沿海县 Coastal County		公共财政收入 Public Revenue of Local Governments		公共财政支出 Public Expenditure of Local Governments	
		2015	2016	2015	2016
山 东	**Shandong**	**8642227**	**8746910**	**10689294**	**10539027**
胶 州	Jiaozhou	800616	920707	929001	1047902
即 墨	Jimo	930493	1050508	1148887	1294772
胶 南	Jiaonan				
垦 利	Kenli	705028	235043	853484	310633
利 津	Lijin	125000	128771	232814	226487
广 饶	Guangrao	424045	409396	513060	499072
长 岛	Changdao	12901	12630	68781	71419
龙 口	Longkou	900517	948167	869280	917182
莱 阳	Laiyang	138376	156710	295820	323378
莱 州	Laizhou	572757	618577	619232	635742
蓬 莱	Penglai	300017	315328	376966	397825
招 远	Zhaoyuan	502200	547388	534618	546015
海 阳	Haiyang	269567	293906	333557	350606
寿 光	Shouguang	900517	946872	948700	971469
昌 邑	Changyi	273326	299086	421786	366415
文 登	Wendeng	492468	495168	579255	609469
荣 成	Rongcheng	672789	688339	1006177	957567
乳 山	Rushan	310358	313298	406036	411441
无 棣	Wudi	197752	250871	332118	380343
沾 化	Zhanhua	113500	116145	219722	221290
广 东	**Guangdong**	**1956123**	**1886095**	**7467577**	**7660897**
南 澳	Nan'ao	21408	21858	89637	86657
台 山	Taishan	247639	243258	431940	452669
恩 平	Enping	98898	98898	248023	248023
遂 溪	Suixi	69476	67277	362115	388142
徐 闻	Xuwen	44664	45338	273562	346890

9-6 续表9 continued

沿海县 Coastal County		公共财政收入 Public Revenue of Local Governments		公共财政支出 Public Expenditure of Local Governments	
		2015	2016	2015	2016
廉 江	Lianjiang	110682	113048	568868	613928
雷 州	Leizhou	58675	50255	553702	606067
吴 川	Wuchuan	68972	66889	316117	369392
电 白	Dianbai	238197	236639	850149	840374
惠 东	Huidong	347174	367751	654386	705906
海 丰	Haifeng	154663	73359	580029	569041
陆 丰	Lufeng	58865	60685	613238	626181
阳 西	Yangxi	73496	66173	335012	230681
阳 东	Yangdong	121618	107081	268308	338765
饶 平	Raoping	76162	80194	445962	427537
揭 东	Jiedong	82088	112956	419962	413116
惠 来	Huilai	83446	74436	456567	397528
广 西	**Guangxi**	**185836**	**190552**	**654998**	**723228**
合 浦	Hepu	67349	65461	402512	458758
东 兴	Dongxing	118487	125091	252486	264470
海 南	**Hainan**	**1669342**	**1739889**	**3717045**	**4140450**
琼 海	Qionghai	182342	154057	430873	437360
儋 州	Danzhou				
文 昌	Wenchang	212983	223784	459660	472820
万 宁	Wanning	164010	155100	414196	434600
东 方	Dongfang	131999	168028	429060	431229
澄 迈	Chengmai	347607	370872	484346	509284
临 高	Lingao	45432	52030	322516	420091
昌 江	Changjiang	87297	100319	276059	308669
乐 东	Ledong	103486	101242	423100	468135
陵 水	Lingshui	394186	414457	477235	658262

9-7 沿海地区一般公共预算收入与支出（2017年）
General Public Budget Revenue and Expenditure by Coastal Regions (2017)

单位：亿元 (100 million yuan)

地 区 Region	地方一般公共预算收入 General Public Budget Revenue	地方一般公共预算支出 General Public Budget Expenditure
合 计 Total	**51072.3**	**75832.7**
天 津 Tianjin	2310.4	3282.5
河 北 Hebei	3233.8	6639.2
辽 宁 Liaoning	2392.8	4879.4
上 海 Shanghai	6642.3	7547.6
江 苏 Jiangsu	8171.5	10621.0
浙 江 Zhejiang	5804.4	7530.3
福 建 Fujian	2809.0	4684.2
山 东 Shandong	6098.6	9258.4
广 东 Guangdong	11320.3	15037.5
广 西 Guangxi	1615.1	4908.6
海 南 Hainan	674.1	1444.0

9-8 沿海地区教育基本情况（2017年）
Basic Conditions of Education by Coastal Regions (2017)

地　区 Region	普通高等学校（机构）数 （所） Number of Regular Schools (Institutions) of Higher Education (unit)	本、专科在校学生数 （人） Number of Enrollment in Normal and Short-cycle Courses (person)	本、专科毕业生数 （人） Number of Students Graduated in Normal and Short-cycle Courses (person)
全国总计 National Total	**2631**	**27535869**	**7358287**
天　津 Tianjin	57	514669	139162
河　北 Hebei	121	1268873	329972
辽　宁 Liaoning	115	980995	268767
上　海 Shanghai	64	514917	134207
江　苏 Jiangsu	167	1767877	489522
浙　江 Zhejiang	107	1002346	276580
福　建 Fujian	89	750987	204417
山　东 Shandong	145	2015345	571220
广　东 Guangdong	151	1925775	511222
广　西 Guangxi	74	866716	210666
海　南 Hainan	19	185538	50370

9-9 沿海地区卫生基本情况（2017年）
Basic Conditions of Public Health by Coastal Regions (2017)

地 区 Region	医疗卫生机构数 （个） Health Care Institutions (unit)	医疗卫生机构床位数 （万张） Number of Beds in Health Care Institutions (10000 beds)	卫生人员 （人） Number of Employed Persons in Health Care Institutions (person)
全国总计 National Total	**986649**	**794.0**	**11748972**
天 津 Tianjin	5539	6.8	129554
河 北 Hebei	80912	39.5	590569
辽 宁 Liaoning	35767	29.9	380915
上 海 Shanghai	5144	13.5	227750
江 苏 Jiangsu	32037	46.9	692473
浙 江 Zhejiang	31979	31.4	555716
福 建 Fujian	27217	18.2	300571
山 东 Shandong	79050	58.5	917894
广 东 Guangdong	49874	49.2	864114
广 西 Guangxi	34008	24.1	404763
海 南 Hainan	5180	4.2	77417

9-10 沿海地区电力消费量
Electricity Consumption by Coastal Region

单位：亿千瓦·时 (100 million kW-h)

地区 Region	2015	2016	2017
天津 Tianjin	800.6	807.9	805.6
河北 Hebei	3175.7	3264.5	3441.7
辽宁 Liaoning	1984.9	2037.4	2135.5
上海 Shanghai	1405.5	1486.0	1526.8
江苏 Jiangsu	5114.7	5458.9	5807.9
浙江 Zhejiang	3553.9	3873.2	4192.6
福建 Fujian	1851.9	1968.6	2112.7
山东 Shandong	5117.0	5390.7	5430.2
广东 Guangdong	5310.7	5610.1	5959.0
广西 Guangxi	1334.3	1359.6	1442.3
海南 Hainan	272.4	287.3	304.8

9-11 沿海地区供水用水情况（2017年）
Water Supply and Water Use by Coastal Regions (2017)

地 区 Region	供水总量 （亿立方米） Water Supply (100 million m^3)	人均用水量 （立方米/人） Per Capita Water Use (m^3/person)
全国总计 National Total	**6043.4**	**435.9**
天 津 Tianjin	27.5	176.3
河 北 Hebei	181.6	242.3
辽 宁 Liaoning	131.1	299.8
上 海 Shanghai	104.8	433.3
江 苏 Jiangsu	591.3	737.8
浙 江 Zhejiang	179.5	319.2
福 建 Fujian	192.0	493.3
山 东 Shandong	209.5	210.0
广 东 Guangdong	433.5	391.1
广 西 Guangxi	284.9	586.0
海 南 Hainan	45.6	494.8

9-12 沿海地区全社会固定资产投资（2017年）
Total Investment in Fixed Assets in the Whole Society by Coastal Regions (2017)

单位：亿元 (100 million yuan)

地 区 Region	全社会固定资产投资 Total Investment in Fixed Assets in the Whole Society	#农、林、牧、渔业 Agriculture, Forestry, Animal Husbandry and Fishery	#交通运输、仓储和邮政业 Transport, Storage and Post
全国总计 National Total	**641238.4**	**26708.0**	**61449.9**
天 津 Tianjin	11288.9	293.0	537.0
河 北 Hebei	33406.8	1873.2	2135.5
辽 宁 Liaoning	6676.7	231.2	602.0
上 海 Shanghai	7246.6	1.6	960.3
江 苏 Jiangsu	53277.0	547.6	2891.0
浙 江 Zhejiang	31696.0	368.9	2967.5
福 建 Fujian	26416.3	1112.6	2808.9
山 东 Shandong	55202.7	1609.6	3955.0
广 东 Guangdong	37761.7	527.8	3759.6
广 西 Guangxi	20499.1	1328.7	2005.6
海 南 Hainan	4244.4	60.6	486.0

9-13 沿海地区按收发货人所在地分货物进出口总额（2017年）

Total Value of Imports and Exports by Location of Importers/Exporters by Coastal Regions (2017)

单位：万美元 (USD 10000)

地 区 Region	进出口 Total	出口 Exports	进口 Imports
全国总计 National Total	**410716427**	**226337133**	**184379294**
天 津 Tianjin	11291916	4356105	6935811
河 北 Hebei	4985554	3135740	1849814
辽 宁 Liaoning	9959508	4486592	5472917
上 海 Shanghai	47619665	19364290	28255375
江 苏 Jiangsu	59077814	36302623	22775191
浙 江 Zhejiang	37790747	28679327	9111420
福 建 Fujian	17102004	10491667	6610337
山 东 Shandong	26455096	14703746	11751349
广 东 Guangdong	100667837	62286837	38381001
广 西 Guangxi	5787866	2809119	2978747
海 南 Hainan	1037405	436591	600815

9-14 沿海地区城镇居民人均可支配收入和人均消费支出（2017年）

Per Capita Disposable Income and Consumption Expenditure of Urban Households by Coastal Regions (2017)

单位：元 (yuan)

地 区 Region	可支配收入 Disposable Income	#工资性收入 Income from Wages and Salaries	消费支出 Consumption Expenditure
全国 National Average	**36396.2**	**22200.9**	**24445.0**
天 津 Tianjin	40277.5	25302.5	30283.6
河 北 Hebei	30547.8	19496.0	20600.3
辽 宁 Liaoning	34993.4	19256.7	25379.4
上 海 Shanghai	62595.7	35995.0	42304.3
江 苏 Jiangsu	43621.8	26298.3	27726.3
浙 江 Zhejiang	51260.7	28817.7	31924.2
福 建 Fujian	39001.4	23886.0	25980.5
山 东 Shandong	36789.4	23431.0	23072.1
广 东 Guangdong	40975.1	30087.3	30197.9
广 西 Guangxi	30502.1	17943.2	18348.6
海 南 Hainan	30817.4	20395.7	20371.9

9-15 沿海地区农村居民人均可支配收入和人均消费支出（2017年）

Per Capita Disposable Income and Consumption Expenditure of Rural Households by Coastal Regions (2017)

单位：元 (yuan)

地　区 Region	可支配收入 Disposable Income	消费支出 Consumption Expenditure
全国 **National Average**	**13432.4**	**10954.5**
天　津 Tianjin	21753.7	16385.9
河　北 Hebei	12880.9	10535.9
辽　宁 Liaoning	13746.8	10787.3
上　海 Shanghai	27825.0	18089.8
江　苏 Jiangsu	19158.0	15611.5
浙　江 Zhejiang	24955.8	18093.4
福　建 Fujian	16334.8	14003.4
山　东 Shandong	15117.5	10342.1
广　东 Guangdong	15779.7	13199.6
广　西 Guangxi	11325.5	9436.6
海　南 Hainan	12901.8	9599.4

9-16 沿海地区年末人口数
Population at Year-end by Coastal Regions

单位：万人 (10000 persons)

地　区 Region	2015	2016	2017
全国总计 National Total	**137462**	**138271**	**139008**
天　津 Tianjin	1547	1562	1557
河　北 Hebei	7425	7470	7520
辽　宁 Liaoning	4382	4378	4369
上　海 Shanghai	2415	2420	2418
江　苏 Jiangsu	7976	7999	8029
浙　江 Zhejiang	5539	5590	5657
福　建 Fujian	3839	3874	3911
山　东 Shandong	9847	9947	10006
广　东 Guangdong	10849	10999	11169
广　西 Guangxi	4796	4838	4885
海　南 Hainan	911	917	926

注：本表数据为年度人口抽样调查推算数据。各地区数据为常住人口口径。

Note: This table is estimated from the annual national sample survey on population.Data by regions are of usual residents.

9-17 沿海城市年末总人口
Total Population at Year-end by Coastal Cities

单位：万人 (10000 persons)

沿海城市	Coastal City	2015	2016	2017
合　计	**Total**	**26683.5**	**26923.0**	**27172.0**
天　津	**Tianjin**	**1547.0**	**1562.0**	**1557.0**
河　北	**Hebei**	**1825.0**	**1838.0**	**1831.0**
唐　山	Tangshan	755.0	760.0	755.0
秦皇岛	Qinhuangdao	295.6	298.0	298.0
沧　州	Cangzhou	774.4	780.0	778.0
辽　宁	**Liaoning**	**1776.5**	**1779.0**	**1765.0**
大　连	Dalian	593.6	596.0	595.0
丹　东	Dandong	238.1	238.0	235.0
锦　州	Jinzhou	302.6	302.0	296.0
营　口	Yingkou	232.6	233.0	232.0
盘　锦	Panjin	129.5	130.0	130.0
葫芦岛	Huludao	280.1	280.0	277.0
上　海	**Shanghai**	**2415.0**	**2420.0**	**2418.0**
江　苏	**Jiangsu**	**2125.4**	**2132.0**	**2123.0**
南　通	Nantong	766.8	767.0	764.0
连云港	Lianyungang	530.6	534.0	533.0
盐　城	Yancheng	828.0	831.0	826.0

注：本表各省数据为沿海城市合计数。

Note: The data for the provinces are the total of coastal cities.

9-17 续表1 continued

沿海城市 Coastal City		2015	2016	2017
浙　江	**Zhejiang**	**3608.9**	**3639.0**	**3719.0**
杭　州	Hangzhou	723.6	736.0	754.0
宁　波	Ningbo	586.6	591.0	597.0
温　州	Wenzhou	811.2	818.0	825.0
嘉　兴	Jiaxing	349.5	352.0	356.0
绍　兴	Shaoxing	443.1	445.0	446.0
舟　山	Zhoushan	97.4	97.0	97.0
台　州	Taizhou	597.5	600.0	604.0
福　建	**Fujian**	**2807.4**	**2848.0**	**2886.0**
福　州	Fuzhou	678.4	687.0	693.0
厦　门	Xiamen	211.2	221.0	231.0
莆　田	Putian	344.3	350.0	355.0
泉　州	Quanzhou	722.5	730.0	742.0
漳　州	Zhangzhou	502.1	508.0	514.0
宁　德	Ningde	348.9	352.0	351.0
山　东	**Shandong**	**3460.6**	**3488.0**	**3514.0**
青　岛	Qingdao	783.1	791.0	803.0
东　营	Dongying	190.6	193.0	195.0
烟　台	Yantai	653.3	655.0	654.0
潍　坊	Weifang	893.7	901.0	908.0
威　海	Weihai	254.8	256.0	256.0
日　照	Rizhao	296.0	300.0	304.0
滨　州	Binzhou	389.1	392.0	394.0

9-17 续表2 continued

沿海城市 Coastal City		2015	2016	2016
广　东	**Guangdong**	**6223.4**	**6312.0**	**6445.0**
广　州	Guangzhou	854.2	870.0	898.0
深　圳	Shenzhen	369.6	385.0	435.0
珠　海	Zhuhai	112.5	115.0	119.0
汕　头	Shantou	550.5	559.0	565.0
江　门	Jiangmen	391.4	394.0	396.0
湛　江	Zhanjiang	823.0	835.0	839.0
茂　名	Maoming	785.8	799.0	804.0
惠　州	Huizhou	357.1	364.0	369.0
汕　尾	Shanwei	359.0	362.0	363.0
阳　江	Yangjiang	292.1	296.0	297.0
东　莞	Dongguan	195.0	201.0	211.0
中　山	Zhongshan	158.7	161.0	170.0
潮　州	Chaozhou	272.8	274.0	276.0
揭　阳	Jieyang	701.7	697.0	703.0
广　西	**Guangxi**	**671.7**	**680.0**	**684.0**
北　海	Beihai	172.0	174.0	175.0
防城港	Fangchenggang	95.6	97.0	98.0
钦　州	Qinzhou	404.1	409.0	411.0
海　南	**Hainan**	**222.6**	**225.0**	**230.0**
海　口	Haikou	164.8	167.0	171.0
三　亚	Sanya	57.8	58.0	59.0
三　沙	Sansha			

9-18 沿海县户籍总人口
Total Population of Household Registration by Coastal Counties

单位：万人 (10000 persons)

沿海县 Coastal County		2015	2016	2017
合　计	**Total**	**8718**	**8720**	**8703**
河　北	**Hebei**	**339**	**319**	**313**
丰　南	Fengnan	53	53	53
滦　南	Luannan	57	58	57
乐　亭	Laoting	45	45	45
昌　黎	Changli	53	56	53
抚　宁	Funing	34	35	33
黄　骅	Huanghua	48	48	48
海　兴	Haixing	49	24	24
辽　宁	**Liaoning**	**657**	**638**	**631**
长　海	Changhai	7	7	7
瓦房店	Wafangdian	100	100	99
普兰店	Pulandian	92	75	73
庄　河	Zhuanghe	90	90	89
东　港	Donggang	61	61	59
凌　海	Linghai	52	51	51
盖　州	Gaizhou	70	70	70
大　洼	Dawa	38	38	39
盘　山	Panshan	28	27	27
绥　中	Suizhong	65	65	64
兴　城	Xingcheng	54	54	53

注：本表各省数据为沿海县合计数。

Note: The data for the provinces are the total of coastal counties.

9-18 续表1 continued

沿海县 Coastal County	2015	2016	2017
江　苏 **Jiangsu**	**1305**	**1306**	**1301**
海　安 Hai'an	94	94	93
如　东 Rudong	104	104	103
启　东 Qidong	112	112	112
海　门 Haimen	100	100	100
赣　榆 Ganyu	120	120	120
东　海 Donghai	123	123	124
灌　云 Guanyun	105	105	104
灌　南 Guannan	82	83	82
响　水 Xiangshui	62	62	62
滨　海 Binhai	122	123	123
射　阳 Sheyang	96	96	96
东　台 Dongtai	113	112	111
大　丰 Dafeng	72	72	71
浙　江 **Zhejiang**	**1496**	**1501**	**1508**
象　山 Xiangshan	55	55	55
宁　海 Ninghai	63	63	63
余　姚 Yuyao	84	84	84
慈　溪 Cixi	105	105	105
奉　化 Fenghua	48	48	48
洞　头 Dongtou	15	15	15
平　阳 Pingyang	88	88	89
苍　南 Cangnan	133	134	135
瑞　安 Rui'an	123	124	125
乐　清 Yueqing	128	130	130
海　盐 Haiyan	38	38	39
海　宁 Haining	68	68	69
平　湖 Pinghu	49	49	50

9-18 续表2 continued

沿海县 Coastal County		2015	2016	2017
越　城	Yuecheng			
柯　桥	Keqiao	65	66	67
上　虞	Shangyu	78	78	78
岱　山	Daishan	19	19	18
嵊　泗	Shengsi	8	8	8
玉　环	Yuhuan	43	43	43
三　门	Sanmen	44	44	45
温　岭	Wenling	122	122	122
临　海	Linhai	120	120	120
福　建	**Fujian**	**1356**	**1367**	**1379**
连　江	Lianjiang	67	67	68
罗　源	Luoyuan	27	27	27
平　潭	Pingtan	43	44	44
福　清	Fuqing	134	136	137
长　乐	Changle	71	73	74
仙　游	Xianyou	114	115	116
惠　安	Hui’an	101	102	102
金　门	Jinmen			
石　狮	Shishi	33	33	34
晋　江	Jinjiang	119	113	115
南　安	Nan’an	159	161	164
云　霄	Yunxiao	45	46	46
漳　浦	Zhangpu	90	92	93
诏　安	Zhao’an	65	66	67
东　山	Dongshan	22	22	22
龙　海	Longhai	86	88	89
霞　浦	Xiapu	54	55	54
福　安	Fu’an	67	67	67
福　鼎	Fuding	59	60	60

9-18 续表3 continued

沿海县 Coastal County	2015	2016	2017
山　东　**Shandong**	**1145**	**1149**	**1149**
胶　州　Jiaozhou	83	84	85
即　墨　Jimo	115	116	117
胶　南　Jiaonan			
垦　利　Kenli	23	23	24
利　津　Lijin	30	28	31
广　饶　Guangrao	52	52	53
长　岛　Changdao	4	4	4
龙　口　Longkou	64	64	64
莱　阳　Laiyang	86	91	87
莱　州　Laizhou	85	85	85
蓬　莱　Penglai	45	45	40
招　远　Zhaoyuan	57	57	57
海　阳　Haiyang	66	64	65
寿　光　Shouguang	107	108	110
昌　邑　Changyi	59	59	59
文　登　Wendeng	58	58	58
荣　成　Rongcheng	67	67	66
乳　山　Rushan	56	56	55
无　棣　Wudi	48	48	49
沾　化　Zhanhua	40	40	40
广　东　**Guangdong**	**1858**	**1876**	**1856**
南　澳　Nan’ao	8	8	8
台　山　Taishan	97	97	97
恩　平　Enping	51	49	50
遂　溪　Suixi	108	110	110
徐　闻　Xuwen	76	77	78

9-18 续表4 continued

沿海县 Coastal County	2015	2016	2017
廉　江 Lianjiang	180	182	182
雷　州 Leizhou	177	181	182
吴　川 Wuchuan	119	120	121
电　白 Dianbai	204	208	193
惠　东 Huidong	87	88	88
海　丰 Haifeng	85	85	86
陆　丰 Lufeng	187	189	190
阳　西 Yangxi	54	54	54
阳　东 Yangdong	50	51	51
饶　平 Raoping	107	107	107
揭　东 Jiedong	111	111	111
惠　来 Huilai	157	159	148
广　西 Guangxi	**122**	**123**	**124**
东　兴 Dongxing	107	108	15
合　浦 Hepu	15	15	109
海　南 Hainan	**440**	**441**	**442**
琼　海 Qionghai	51	51	52
儋　州 Danzhou			
文　昌 Wenchang	57	60	60
万　宁 Wanning	64	62	62
东　方 Dongfang	44	45	45
澄　迈 Chengmai	58	56	56
临　高 Lingao	49	50	50
昌　江 Changjiang	25	26	25
乐　东 Ledong	53	53	54
陵　水 Lingshui	39	38	38

9-19 沿海地区就业人员情况
Employed Persons by Coastal Regions

单位：万人 (10000 persons)

地 区 Region	2015	2016	2017
合 计 Total	**36368.9**	**36434.0**	**36445.7**
天 津 Tianjin	896.8	902.4	894.8
河 北 Hebei	4212.5	4224.0	4206.7
辽 宁 Liaoning	2409.9	2301.2	2284.7
上 海 Shanghai	1361.5	1365.2	1372.7
江 苏 Jiangsu	4758.5	4756.2	4757.8
浙 江 Zhejiang	3733.7	3760.0	3796.0
福 建 Fujian	2768.4	2797.0	2805.7
山 东 Shandong	6632.5	6649.7	6560.6
广 东 Guangdong	6219.3	6279.2	6340.8
广 西 Guangxi	2820.0	2841.0	2842.0
海 南 Hainan	555.8	558.1	583.9

9-20 沿海城市城镇单位就业人员情况
Employed Persons in Urban Units by Coastal Cities

单位：万人 (10000 persons)

沿海城市 Coastal City		2015	2016	2017
合 计	**Total**	**5445.1**	**5385.0**	**5312.6**
天 津	**Tianjin**	**294.8**	**286.0**	**269.5**
河 北	**Hebei**	**174.8**	**172.5**	**148.9**
唐 山	Tangshan	89.4	88.1	76.1
秦皇岛	Qinhuangdao	32.8	32.5	29.3
沧 州	Cangzhou	52.6	51.9	43.5
辽 宁	**Liaoning**	**266.8**	**251.7**	**228.1**
大 连	Dalian	113.7	107.8	97.3
丹 东	Dandong	25.8	23.0	19.7
锦 州	Jinzhou	31.7	29.3	24.1
营 口	Yingkou	24.4	25.7	25.2
盘 锦	Panjin	46.4	44.2	41.7
葫芦岛	Huludao	24.8	21.7	20.1
上 海	**Shanghai**	**637.2**	**627.8**	**632.3**
江 苏	**Jiangsu**	**346.7**	**340.2**	**338.8**
南 通	Nantong	209.8	205.3	209.7
连云港	Lianyungang	47.7	47.5	45.7
盐 城	Yancheng	89.2	87.4	83.4
浙 江	**Zhejiang**	**929.2**	**905.0**	**886.3**
杭 州	Hangzhou	288.6	290.1	287.0
宁 波	Ningbo	166.8	151.9	161.3
温 州	Wenzhou	106.6	104.6	119.1
嘉 兴	Jiaxing	80.9	80.5	79.1
绍 兴	Shaoxing	138.6	136.8	120.2
舟 山	Zhoushan	46.6	47.0	18.4
台 州	Taizhou	101.1	94.1	101.2
福 建	**Fujian**	**579.6**	**585.1**	**587.2**
福 州	Fuzhou	156.3	156.8	158.8
厦 门	Xiamen	136.8	139.2	146.0

注：本表各省数据为沿海城市合计数。

Note: The data for the provinces are the total of coastal cities.

9-20 续表 continued

沿海城市 Coastal City	2015	2016	2017
莆田 Putian	49.7	52.0	54.1
泉州 Quanzhou	150.6	149.6	139.9
漳州 Zhangzhou	55.2	56.0	56.7
宁德 Ningde	31.0	31.5	31.7
山东 Shandong	**527.3**	**518.2**	**505.6**
青岛 Qingdao	150.1	145.4	145.9
东营 Dongying	43.9	43.2	40.0
烟台 Yantai	105.2	103.5	99.8
潍坊 Weifang	87.0	85.7	83.9
威海 Weihai	57.6	58.7	57.5
日照 Rizhao	31.0	31.1	31.6
滨州 Binzhou	52.5	50.6	46.9
广东 Guangdong	**1580.6**	**1588.0**	**1605.4**
广州 Guangzhou	320.3	325.2	329.2
深圳 Shenzhen	460.0	456.2	463.8
珠海 Zhuhai	74.3	73.1	76.2
汕头 Shantou	54.7	57.5	59.6
江门 Jiangmen	58.3	59.5	56.7
湛江 Zhanjiang	51.0	52.3	50.9
茂名 Maoming	45.7	46.5	49.4
惠州 Huizhou	91.9	96.1	98.8
汕尾 Shanwei	24.0	23.8	20.4
阳江 Yangjiang	23.6	24.2	23.7
东莞 Dongguan	232.3	231.2	242.4
中山 Zhongshan	82.9	80.9	77.9
潮州 Chaozhou	19.8	20.2	18.5
揭阳 Jieyang	41.8	41.3	37.9
广西 Guangxi	**46.3**	**46.0**	**44.4**
北海 Beihai	14.7	14.5	14.4
防城港 Fangchenggang	10.8	9.7	9.3
钦州 Qinzhou	20.8	21.8	20.7
海南 Hainan	**61.8**	**64.5**	**66.1**
海口 Haikou	49.2	51.4	52.2
三亚 Sanya	12.6	13.1	13.9
三沙 Sansha			

主要统计指标解释

1. 国内(或地区)生产总值 指一个国家（或地区）所有常住单位在一定时期内生产活动的最终成果。国内生产总值有三种表现形态，即价值形态、收入形态和产品形态。从价值形态看，它是所有常住单位在一定时期内生产的全部货物和服务价值与同期投入的全部非固定资产货物和服务价值的差额，即所有常住单位的增加值之和；从收入形态看，它是所有常住单位在一定时期内创造的各项收入之和，包括劳动者报酬、生产税净额、固定资产折旧和营业盈余；从产品形态看，它是所有常住单位在一定时期内最终使用的货物和服务价值与货物和服务净出口价值之和。在实际核算中，国内生产总值有三种计算方法，即生产法、收入法和支出法。三种方法分别从不同的方面反映国内生产总值及其构成。

2. 三次产业 三次产业的划分是世界上较为常用的产业结构分类，但各国的划分不尽一致。根据《国民经济行业分类》（GB/T 4754—2011）和《三次产业划分规定》，我国的三次产业划分是：

第一产业是指农、林、牧、渔业（不含农、林、牧、渔服务业）。

第二产业是指采矿业（不含开采辅助活动），制造业（不含金属制品、机械和设备修理业），电力、热力、燃气及水生产和供应业，建筑业。

第三产业即服务业，是指除第一产业、第二产业以外的其他行业。

3. 增加值 是指各行各业生产经营和劳务活动的最终成果，采用生产法和收入法两种方法计算。

生产法是从货物和服务活动在生产过程中形成的总产品入手，剔除生产过程中投入的中间产品价值，得到新增价值的方法。

收入法又称分配法。按收入法计算国内生产总值是从生产过程创造的收入的角度对常驻单位的生产活动成果进行核算；按照此法计算，增加值由劳动者报酬、固定资产折旧、生产税净额和营业盈余四个部分组成。

4. 财政收入 指国家财政参与社会产品分配所取得的收入，是实现国家职能的财力保证。财政收入所包括的内容几经变化，目前主要包括：

（1）各项税收 包括增值税、营业税、消费税、土地增值税、城市维护建设税、资源税、城市土地使用税、企业所得税、个人所得税、关税、证券交易印花税、车辆购置税、农牧业税和耕地占用税等；

（2）专项收入 包括排污费收入、城市水资源费收入、矿产资源补偿费收入、教育费附加收入等；

（3）其他收入 包括利息收入、基本建设贷款归还收入、基本建设收入、捐赠收入等；

（4）国有企业亏损补贴 此项为负收入，冲减财政收入。主要包括对工业企业、商业企业、粮食企业的补贴。

5. 财政支出 国家财政将筹集起来的资金进行分配使用，以满足经济建设和各项事业的需要。

6. 普通高等学校 指通过国家普通高等教育招生考试，招收高中毕业生为主要培养对象，实施高等学历教育的全日制大学、独立设置的学院、独立学院和高等专科学校、高等职业学校及其他机构。

大学、独立设置的学院主要实施本科及本科层次以上的教育。独立学院主要实施本科层次的教育。高等专科学校、高等职业学校实施专科层次的教育。其他机构是指承担国家普通招生计划任务但不计校数的机构，包括普通高等学校分校、大专班等。

7. 医疗卫生机构 指从卫生(卫生计生)行政部门取得《医疗机构执业许可证》《计划生育技术服务许可证》，或从民政、工商行政、机构编制管理部门取得法人单位登记证书，为社会提供医疗服务、公共卫生服务或从事医学科研和医学在职培训等工作的单位。医疗卫生机构包括医院、基层医疗卫生机构、专业公共卫生机构、其他医疗卫生机构。

8. 供水总量 指各种水源为用水户提供的包括输水损失在内的毛水量。

9. 全社会固定资产投资 是以货币形式表现的在一定时期内全社会建造和购置固定资产的工作量以及与此有关的费用的总称。该指标是反映固定资产投资规模、结构和发展速度的综合性指标，又是观察工程进度和考核投资效果的重要依据。全社会固定资产投资按登记注册类型可分为国有、集体、联营、股份制、私营和个体、港澳台商、外商、其他等。

10. 货物进出口总额 指实际进出我国关境的货物总金额。包括对外贸易实际进出口货物，来料加工装配进出口货物，国家间、联合国及国际组织无偿援助物资和赠送品，华侨、港澳台同胞和外籍华人捐赠品，租赁期满归承租人所有的租赁货物，进料加工进出口货物，边境地方贸易及边境地区小额贸易进出口货物，中外合资企业、中外合作经营企业、外商独资经营企业进出口货物和公用物品，到、离岸价格在规定限额以上的进出口货样和广告品(无商业价值、无使用价值和免费提供出口的除外)，从保税仓库提取在中国境内销售的进口货物，以及其他进出口货物。该指标可以观察一个国家在货物贸易方面的总规模。我国规定出口货物按离岸价格统计，进口货物按到岸价格统计。

11. 商品收发货人所在地进、出口额 指在所在地海关注册登记的有进出口经营权的企业实际进、出口额。

12. 人口数 指一定时点、一定地区范围内有生命的个人总和。

年度统计的年末人口数指每年 12 月 31 日 24 时的人口数。年度统计的全国人口总数内未包括香港、澳门特别行政区和台湾省以及海外华侨人数。

13. 就业人员 指在 16 周岁及以上，从事一定社会劳动并取得劳动报酬或经营收入的人员。这一指标反映了一定时期内全部劳动力资源的实际利用情况，是研究我国基本国情国力的重要指标。

Explanatory Notes on Main Statistical Indicators

1. Gross Domestic Product (GDP) refers to the final products produced by all resident units in a country (or a region) during a certain period of time. Gross domestic product is expressed in three different perspectives, namely value, income, and products respectively. GDP in its value perspective refers to the balance of total value of all goods and services produced by all resident units during a certain period of time, minus the total value of input of goods and services of the nature of non-fixed assets; in other words, it is the sum of the value-added of all resident units. GDP from the perspective of income refers to the sum of all kinds of revenue, including Compensation of Employees, Net Taxes on Production, Depreciation of Fixed Assets, and Operating Surplus. GDP from the perspective of products refers to the value of all goods and services for final demand by all resident units plus the net exports of goods and services during a given period of time. In the practice of national accounting, gross domestic product is calculated from three approaches, namely production approach, income approach and expenditure approach, which reflect gross domestic product and its composition from different angles.

2. Three Industries Classification of economic activities into three strata of industry is a common practice in the world, although the grouping varies to some extent from country to country. In China, according to Industrial classification for National Economic Activities (GB/T 4754—2011) and Dividing Basis of Three Industries, economic activities are categorized into the following three strata of industry:

Primary industry refers to agriculture, forestry, animal husbandry and fishery industries (not including services in support of agriculture, forestry, animal husbandry and fishery industries).

Secondary industry refers to mining and quarrying(not including support activities for mining), manufacturing(not including repair service of metal products, machinery and equipment), production and supply of electricity, heat, gas and water, and construction.

Tertiary industry refers to all other economic activities not included in the primary or secondary industries.

3. Added Value refers to the final result of production operation and labor activities of all trades and professions, which is calculated by using the methods of production and income.

Production Method: refers to the method whereby to get the newly added value by proceeding from the gross product of goods and service activities occurring in the course of production and then rejecting the value of intermediate product input in the course of production.

Income Method: is also called the distribution method. The calculation of the gross domestic product (GDP) by the income method is the accounting of the result of productive activities of permanent units from the angle of the income created in the course of production. According to this method, the added value is composed of the payment for laborers, depreciation for fixed assets, net tax on production and business surplus.

4. Government Revenue refers to the income obtained by the government finance through participating in the distribution of social products. It is the financial guarantee to ensure government functioning. The contents of government revenue have changed several times. Now it includes the following main items:

(1) Various tax revenues, including value added tax, business tax, consumption tax, land value-added tax, tax on city maintenance and construction, resources tax, tax on use of urban land, enterprise income tax, personal income tax, tariff, stamp tax on security transactions, tax on purchase of motor vehicles, tax on agriculture and animal husbandry and tax on occupancy of cultivated land, etc.

(2) Special revenues, including revenues from the fee on sewage treatment, fee on urban water resources, fee for the compensation of mineral resources and extra-charges for education, etc.

(3) Other revenues, including revenues from interest, repayment of capital construction loan, capital construction projects, and donations and grants.

(4) Subsidies for the losses of State-owned enterprises. This is an item of negative revenue, counteracting revenues and consisting of subsidies to industrial, commercial and grain purchasing and supply enterprises.

5. Government Expenditure refers to the distribution and use of the funds which the government finance has raised, so as to meet the needs of economic construction and various causes.

6. Regular Institutions of Higher Learning refers to educational establishments recruiting graduates from senior secondary schools as the main target through National Matriculation TEST. They include full-time universities, independently established colleges, colleges, and institutions of higher professional education, institutions of higher vocational education and others.

Universities and independently established colleges primarily provide undergraduate and above courses; colleges mainly impart undergraduate courses, institutions of higher professional education and institutions of higher vocational education primarily provide professional trainings; and others refer to educational establishments, which are responsible for enrolling higher education students under the State Plan but not enumerated in the total number of schools, including: branch schools of universities and colleges and junior colleges.

7. Medical and Health Care Institution refers to the units which have been qualified the Certification of Health Care Institution, certification of family planning technical service by the administration of public health (family planning), or qualified the Certification of Corporate Unit by the civil affairs, administration for industry and commerce, commission office for public sector reform, and engaging in medical health care services, public health services, or medicine research and on-job training, etc., including: hospitals, health care institutions at grass-root level, specialized public health institutions, and other medical and health care institutions.

8. Water Supply refers to gross water of various sources supplied to consumers, including losses during distribution.

9. Total Investment in Fixed Assets in the Whole Country refers to the volume of activities in construction and purchases of fixed assets of the whole country and related fees, expressed in monetary terms during the reference period. It is a comprehensive indicator which shows the size, structure and growth of the investment in fixed assets, providing a basis for observing the progress of construction projects and evaluating results of investment. Total investment in fixed assets in the whole country includes, by type of ownership, the investment by State-owned units, collective-owned units, joint ownership units, share-holding units, private units, individuals as well as investments by entrepreneurs from Hong Kong, Macao and Taiwan, foreign investors and others.

10. Total Import and Export of Goods refers to the real value of commodities imported and exported across the border of China. They include the actual imports and exports through foreign trade, imported and exported goods under the processing and assembling trades and materials, supplies and gifts as aid given gratis between governments and by the United Nations and other international organizations, and contributions donated by overseas Chinese, compatriots in Hong Kong and Macao and Chinese with foreign citizenship, leasing commodities owned by tenant at the expiration of leasing period, the imported and exported commodities processed with imported materials, commodities trading in border areas, the imported and exported commodities and articles for public use of the Sino-foreign joint ventures, cooperative enterprises and ventures with sole foreign investment. Also included is import or export of samples and advertising goods for which CIF or FOB value are beyond the permitted ceiling (excluding goods of no trading or use value and free commodities for export), imported goods sold in China from bonded warehouses and other imported or exported goods. The indicator of the total imports and exports at customs can be used to observe the total size of external trade in a country. In accordance with the stipulation of the Chinese government, imports are calculated at CIF, while exports are calculated at FOB.

11. Import or Export Value by Location of China's Foreign Trade Managing Units refers to actual value of imports and exports carried out by corporations which have been registered by the local Customs house and are vested with right to run import export business.

12. Total Population refers to the total number of people alive at a certain point of time within a given area.

The annual statistics on total population is taken at midnight, the 31st of December, not including residents in Taiwan province, Hong Kong and Macao and overseas Chinese.

13. Employed Persons refers to persons aged 16 and over who are engaged in gainful employment and thus receive remuneration payment or earn business income. This indicator reflects the actual utilization of total labour force during a certain period of time and is often used for the research on China's economic situation and national power.

10

世界海洋经济统计资料（部分）
World′s Marine Economic Statistics Data (Part)

10-1 世界海洋面积（2010年）
World Ocean Area (2010)

区 域 Region	海洋面积 （平方千米） Ocean Area (km^2)	占世界海洋面积的比重（%） Proportion in the World Ocean Area (%)	占地球表面面积的比重（%） Proportion in the Earth's Surface Area (%)
合 计 **Total**	**361000000**	**100.0**	**70.8**
太 平 洋 Pacific Ocean	178334000	49.4	35.0
大 西 洋 Atlantic Ocean	91694000	25.4	18.0
印 度 洋 Indian Ocean	76171000	21.1	14.9
北 冰 洋 Arctic Ocean	14801000	4.1	2.9

注：数据来源于《2011国际统计年鉴》。

Note: The data come from *International Statistical Yearbook 2011* .

10-2 主要沿海国家（地区）海岸线长度
Length of Coastline of Major Coastal Countries (Areas)

单位：千米 (km)

国家或地区 Country or Area	海岸线长度 Length of Coastline
中国 China	32000
美国 United States	22680
日本 Japan	30000
德国 Germany	1300
英国 United Kingdom	11450
法国 France	3000
意大利 Italy	7000
加拿大 Canada	20000
澳大利亚 Australia	20125
俄罗斯 Russia	34000
波兰 Poland	491
印度 India	6083
印度尼西亚 Indonesia	35000
菲律宾 Philippines	18533
泰国 Thailand	3219
马来西亚 Malaysia	4675
新加坡 Singapore	193
缅甸 Myanmar	3060
孟加拉国 Bangladesh	580
土耳其 Turkey	7200
韩国 Korea, Rep.	2413
埃及 Egypt	2450
墨西哥 Mexico	9330
巴西 Brazil	7400
阿根廷 Argentina	4989

10-3 主要沿海国家（地区）国土面积和人口（2017年）
Surface Area and Population of Major Coastal Countries (Areas) (2017)

国家或地区 Country or Area	国土面积（万平方千米） Area of Territory (10 000 km^2)	年中人口（万人） Mid-year Population (10 000 persons)	人口密度 （人/平方千米） Population Density (persons per km^2)
世界 World	**13432.5**	**753036.0**	**58.0**
中国 China	960.0	138640.0	148.0
文莱 Brunei Darsm	0.6	43.0	81.0
柬埔寨 Cambodia	18.1	1601.0	91.0
印度 India	298.0	133918.0	450.0
印度尼西亚 Indonesia	191.1	26399.0	146.0
日本 Japan	37.8	12679.0	348.0
韩国 Korea, Rep.	10.0	5147.0	528.0
马来西亚 Malaysia	33.1	3162.0	96.0
缅甸 Myanmar	67.7	5337.0	82.0
菲律宾 Philippines	30.0	10492.0	352.0
新加坡 Singapore	0.1	561.0	7916.0
泰国 Thailand	51.3	6904.0	135.0
越南 Viet Nam	33.1	9554.0	308.0
埃及 Egypt	100.2	9755.0	98.0
南非 South Africa	121.9	5672.0	47.0
加拿大 Canada	998.5	3671.0	4.0
墨西哥 Mexico	196.4	12916.0	66.0
美国 United States	983.2	32572.0	36.0
阿根廷 Argentina	278.0	4427.0	16.0
巴西 Brazil	851.6	20929.0	25.0
法国 France	54.9	6712.0	123.0
德国 Germany	35.7	8270.0	237.0
意大利 Italy	30.1	6055.0	206.0
俄罗斯 Russia	1709.8	14450.0	9.0
西班牙 Spain	50.6	4657.0	93.0
土耳其 Turkey	78.5	8075.0	105.0
乌克兰 Ukraine	60.4	4483.0	77.0
英国 United Kingdom	24.4	6602.0	273.0
澳大利亚 Australia	774.1	2460.0	3.0
新西兰 New Zealand	26.8	479.0	18.0

注：数据来源于世界银行WDI数据库。

Note: The data come from *World Bank WDI Database.*

10-4 主要沿海国家（地区）国内生产总值（2017年）
Gross Domestic Product of Major Coastal Countries (Areas) (2017)

国家或地区 Country or Area	国内生产总值 （亿美元） GDP (100 million USD)	人均国内生产总值 （美元） GDP per Captia (current USD)	国内生产总值增长率 （%） Growth Rate of GDP (%)
世界 **World**	806838	10715	3.2
中国 China	122377	8827	6.9
中国香港 Hong Kong, China	3414	46194	3.8
文莱 Brunei Darsm	121	28291	1.3
柬埔寨 Cambodia	222	1384	6.8
印度 India	25975	1940	6.6
印度尼西亚 Indonesia	10155	3847	5.1
日本 Japan	48721	38428	1.7
韩国 Korea, Rep.	15308	29743	3.1
马来西亚 Malaysia	3145	9945	5.9
菲律宾 Philippines	3136	2989	6.7
新加坡 Singapore	3239	57714	3.6
泰国 Thailand	4552	6594	3.9
越南 Viet Nam	2239	2343	6.8
埃及 Egypt	2354	2413	4.2
南非 South Africa	3494	6161	1.3
加拿大 Canada	16530	45032	3.0
墨西哥 Mexico	11499	8903	2.0
美国 United States	193906	59532	2.3
阿根廷 Argentina	6376	14402	2.9
巴西 Brazil	20555	9821	1.0
法国 France	25825	38477	1.8
德国 Germany	36774	44470	2.2
意大利 Italy	19348	31953	1.5
荷兰 Netherlands	8262	48223	3.2
波兰 Poland	5245	13812	4.6
俄罗斯 Russian	15775	10743	1.5
西班牙 Spain	13113	28157	3.1
土耳其 Turkey	8511	10541	7.4
乌克兰 Ukraine	1122	2640	2.5
英国 United Kingdom	26224	39720	1.8
澳大利亚 Australia	13234	53800	2.0

注：数据来源于世界银行WDI数据库。

Note: The data come from *World Bank WDI Database.*

10-5 主要沿海国家（地区）国内生产总值产业构成（2017年）

Industrial Composition of GDP of Major Coastal Countries (Areas) by Industry (2017)

单位：% (%)

国家或地区 Country or Area	国内生产总值产业构成（%） Industrial Structure of GDP (%)		
	第一产业 Primary Industry	第二产业 Secondary Industry	第三产业 Tertiary Industry
世界 World	**3.5** ①	**25.4** ①	**65.1** ①
中国 China	7.9	40.5	51.6
中国香港 Hong Kong,China	0.1 ①	7.5 ①	89.5 ①
文莱 Brunei Darsm	1.1	59.7	40.9
柬埔寨 Cambodia	23.4	30.9	39.7
印度 India	15.5	26.2	48.9
印度尼西亚 Indonesia	13.1	39.4	43.6
日本 Japan	1.2 ①	29.3 ①	68.8 ①
韩国 Korea, Rep.	2.0	35.9	52.8
马来西亚 Malaysia	8.8	38.8	51.0
缅甸 Myanmar	26.2	31.6	42.2
菲律宾 Philippines	9.7	30.5	59.9
新加坡 Singapore		23.2	70.4
斯里兰卡 Sri Lanka	7.7	27.2	55.8
泰国 Thailand	8.7	35.0	56.3
越南 Viet Nam	15.3	33.3	40.9 ①
埃及 Egypt	11.5	33.8	55.2 ①
南非 South Africa	2.3	25.9	61.5
加拿大 Canada	1.4 ②	27.5 ②	64.7 ②
墨西哥 Mexico	3.4	29.9	60.9
美国 United States	1.0 ①	18.9 ①	77.0 ①
阿根廷 Argentina	5.6	21.7	56.9
巴西 Brazil	4.6	18.5	63.1
法国 France	1.5	17.4	70.2
德国 Germany	0.6	27.6	61.9
意大利 Italy	1.9	21.4	66.3
荷兰 Netherlands	1.9	17.5	70.4
俄罗斯 Russia	4.0	30.0	56.2
西班牙 Spain	2.6	21.6	66.4
土耳其 Turkey	6.1	29.2	53.3
英国 United Kingdom	0.5	18.6	70.1
澳大利亚 Australia	2.8	23.0	67.0
新西兰 New Zealand	5.5 ③	20.4 ③	65.6 ③

注：①2016年数据。②2014年数据。③2015年数据。
数据来源于世界银行WDI数据库。

Note: ①Data for 2016. ②Data for 2014. ③Data for 2015.
The data come from *World Bank WDI Database*.

10-6 主要沿海国家（地区）就业人数
Employment in the Major Coastal Countries (Areas)

单位：万人 (10000 persons)

国家或地区 Country or Area	2000	2005	2010	2014	2015	2016
中国 China	72085	74647	76105	77253	77451	77603
中国香港 Hong Kong,China	321	334	347	375	378	
印度 India	33020	37199	37429			
印度尼西亚 Indonesia	8984	9536	10959	11463	11482	
以色列 Israel	222	249	294	356	364	
日本 Japan	6446	6356	6298	6351	6376	6440
韩国 Korea, Rep.	2116	2286	2383	2560	2594	2641
马来西亚 Malaysia	932	1005	1178	1353	1407	1416
菲律宾 Philippines	2745	3231	3604	3865	3874	4084
新加坡 Singapore	209	165	306	210	215	
斯里兰卡 Sri Lanka	631	752	771	842	855	
泰国 Thailand	3300	3630	3804	3842	3802	3769
越南 Viet Nam	3837	4253	4949	5274	5291	5330
埃及 Egypt	1720	1934	2383	2430	2478	2537
南非 South Africa	1224	1230	1379	1532	1574	
加拿大 Canada	1476	1612	1696	1780	1795	
墨西哥 Mexico	3373	4079	4612	4737	5031	5160
美国 United States	13689	14173	13906	14631	14883	15144
阿根廷 Argentina	826	964	1516	1569		
巴西 Brazil	6563	1955	2202	9945	2308	8951
委内瑞拉 Venezuela	896	1073	1207	1319	1321	
法国 France	2312	2498	2573	2640	2642	2658
德国 Germany	3632	3636	3799	3987	4021	4127
意大利 Italy	2093	2241	2253	2228	2246	2276
荷兰 Netherlands	786	811	837	824	832	843
波兰 Poland	1452	1412	1547	1586	1608	1620
俄罗斯 Russian	6507	6817	6980	7154	7232	
西班牙 Spain	1544	1921	1872	1734	1787	1834
土耳其 Turkey	2158	2205	2259	2593	2662	2722
乌克兰 Ukraine	2018	2068	2027	1807	1644	1628
英国 United Kingdom	2726	2874	2912	3067	3120	3163
澳大利亚 Australia	886	985	1099	1154	1177	1195
新西兰 New Zealand	178	208	218	231	236	

注：数据来源于联合国ILO数据库。

Note: The data come from *ILO Database*.

10-7 主要沿海国家（地区）鱼类产量（2016年）
Fish Yields of Major Coastal Countries (Areas) (2016)

单位：万吨 (10000 t)

国家或地区 Country or Area	鱼类产量 Output of Total Fishes	海域鱼类产量 Ocean Area
中国 China	4217.2	1230.6
印度 India	939.9	292.7
印度尼西亚 Indonesia	997.9	614.1
缅甸 Myanmar	298.9	115.6
俄罗斯 Russian	470.3	426.4
美国 United States	422.2	402.2
越南 Viet Nam	486.2	238.5
日本 Japan	296.8	291.8
孟加拉国 Bangladesh	363.2	69.2
菲律宾 Philippines	252.1	212.2
泰国 Thailand	167.2	108.7
马来西亚 Malaysia	150.6	139.8
巴西 Brazil	114.5	41.8
墨西哥 Mexico	128.1	101.1
埃及 Egypt	167.7	8.5
韩国 Korea, Rep.	114.2	111.4
西班牙 Spain	92.2	89.9
尼日利亚 Nigeria	100.2	31.8
南非 South Africa	60.2	59.9
英国 United Kingdom	72.9	71.7
柬埔寨 Cambodia	76.7	8.9
土耳其 Turkey	54.5	41.2
阿根廷 Argentina	48.2	46.0
加拿大 Canada	61.1	57.0
斯里兰卡 Sri Lanka	50.1	40.3
新西兰 New Zealand	38.3	38.1
法国 France	46.3	42.7
荷兰 Netherlands		

注：数据来源于联合国FAO数据库。

Note: The data come from *FAO Database*.

10-8 国家保护区面积和鱼类濒危物种（2017年）
Area of National Nature Reserves and Endangered Species of Fish (2017)

国家或地区 Country or Area	国家保护区[①] National Protected		鱼类濒危物种（种）Endangered Species of Fish (number)
	陆地保护区面积占陆地面积比重 Terrestrial Protected Areas (% of Total Land Area)	海洋保护区面积占领海面积比重 Marine Protected Areas (% of Territorial Waters)	
中国 China	17.1	3.8	134
中国香港 Hong Kong, China	41.9		15
孟加拉国 Bangladesh	4.6	5.4	29
文莱 Brunei Darsm	46.9	0.2	14
柬埔寨 Cambodia	26.0	0.2	49
印度 India	6.0	0.2	228
印度尼西亚 Indonesia	11.9	2.9	163
伊朗 Iran	8.6	0.8	46
日本 Japan	19.4	0.5	77
韩国 Korea, Rep.	11.2	1.6	27
马来西亚 Malaysia	19.1	1.4	87
菲律宾 Philippines	15.3	1.2	93
新加坡 Singapore	5.6		29
斯里兰卡 Sri Lanka	29.9	0.1	57
泰国 Thailand	18.8	1.9	107
越南 Viet Nam	7.6	0.6	82
埃及 Egypt	13.1	5.0	54
尼日利亚 Nigeria	13.9	0.0	72
南非 South Africa	14.1	12.1	120
加拿大 Canada	9.7	0.9	43
墨西哥 Mexico	14.3	2.3	181
美国 United States	13.0	41.1	251
阿根廷 Argentina	8.9	4.0	41
巴西 Brazil	28.9	1.7	90
委内瑞拉 Venezuela	54.1	3.5	44
法国 France	26.0	26.2	52
德国 Germany	37.7	45.4	24
意大利 Italy	21.5	8.8	51
荷兰 Netherlands	11.3	21.5	15
波兰 Poland	39.6	22.6	8
俄罗斯 Russia	9.7	3.0	39
西班牙 Spain	28.0	8.7	80
土耳其 Turkey	0.2	0.1	131
乌克兰 Ukraine	4.0	3.4	24
英国 United Kingdom	28.2	20.2	47
澳大利亚 Australia	17.0	40.7	123
新西兰 New Zealand	32.6	30.3	35

注：①2016年数据。

数据来源于世界银行WDI数据库。

Note: ①Data refer to 2016.

The data come from *World Bank WDI Database.*

10-9 主要沿海国家（地区）风力发电量
Wind-Power Capacities of Major Coastal Countries (Areas)

单位：百万千瓦·时　　　　(million kilowatt-hours)

国家或地区 Country or Area	2007	2008	2009	2010	2011	2013	2014	2015
中国 China		13079	46096	44622	70331	141197	156078	185766
伊朗 Iran	143	196	224			376	358	221
以色列 Israel	1	9	9	8		6	6	7
日本 Japan	2624	2623	2949	3962	7	5201	5038	5160
韩国 Korea, Rep	376	436	685	817	863	1149	1146	1201
马来西亚 Malaysia								
巴基斯坦 Pakistan							397	1549
菲律宾 Philippines	59	61	64	62	88	66	152	748
新加坡 Singapore								
斯里兰卡 Sri Lanka	2	3	3	53	92	236	273	344
泰国 Thailand				3	3	305	305	329
越南 Viet Nam						90	300	261
埃及 Egypt	831	931	1133	1498	1525	1332	1315	1345
尼日利亚 Nigeria								
南非 South Africa	32	32	32	32	103	37	1070	2270
加拿大 Canada	3024	3819	4573	9557	10187	11594	22538	26446
墨西哥 Mexico	262	269	596	1239	1648	4185	6426	8745
美国 United States	34603	55696	74226	95148	120854	169713	183892	192992
阿根廷 Argentina	61	42	36	25	26	461	730	599
巴西 Brazil					2705	6579	12211	21626
委内瑞拉 Venezuela								
法国 France	4052	5689	7891	9969	12052	16033	17249	21249
德国 Germany	39713	40574	38639	37793	48883	51708	57357	79206
意大利 Italy	4034	4861	6543	9126	9856	14897	15178	14844
荷兰 Netherlands	3438	4260	4581	3993	5100	5627	5797	7550
波兰 Poland	522	837	1077	1664	3205	6004	7676	10858
俄罗斯 Russia	7	5	4	4	5	5	96	148
西班牙 Spain	27509	32203	37773	44165	42918	53903	52013	49325
土耳其 Turkey	355	847	1495	2916	4723	7557	8520	11652
乌克兰 Ukraine	45	45	43	50	89	639	1130	1084
英国 United Kingdom	5274	7097	9304	10183	15509	28434	32015	40310
澳大利亚 Australia	2611	3941	3806	4798	5807	7328	10252	11467
新西兰 New Zealand	937	1057	1471	1634	1952	2020	2214	2356

注：数据来源于联合国ESD数据库。

Note: The data come from *UN ESD Database*.

10-10 主要沿海国家（地区）捕捞产量
Fishing Yields in the Major Coastal Countries (Areas)

单位：吨 (t)

国家或地区 Country or Area	2015	2016
世界总计 World Total	**92630460**	**90909868**
中国 China	17591299	17564280
秘鲁 Peru	4824050	3796978
印度尼西亚 Indonesia	6485320	6542258
美国 United States	5038791	4919741
印度 India	4843388	5061756 ①
俄罗斯 Russia	4457138	4759331
日本 Japan	3460168 ①	3195558 ①
缅甸 Myanmar	1953510 ①	2072390 ①
智利 Chile	1786633	1499531
越南 Viet Nam	2757314	2785940
菲律宾 Philippines	2151502	2024828
挪威 Norway	2293698	2033953
泰国 Thailand	1693050	1530583
韩国 Korea, Rep.	1648993	1386446
孟加拉国 Bangladesh	1623837	1674770
墨西哥 Mexico	1467203	1510754
马来西亚 Malaysia	1491974	1580291
冰岛 Iceland	1317349	1067191
西班牙 Spain	973240	911638

注：捕捞品种包括鱼类、甲壳类、软体类等水生动物。①为联合国粮农组织估算值。

数据来源于《渔业和水产养殖统计年鉴》，联合国粮农组织，2016年。

Note: The fished species include fish, crustacea, mollusc and other aquatic animals.

① It is estimated by FAO from available sources of information or calculation.

The data come from *Fishery and Aquaculture Statistical Yearbook,* FAO, 2016.

10-10 续表1 continued

国家或地区 Country or Area	2015	2016
摩洛哥 Morocco	1364643	1447020
中国台湾 Taiwan, China	987873	750110
加拿大 Canada	851119	861997
巴西 Brazil	700000	705000 ①
阿根廷 Argentina	814300	755226
南非 South Africa	564969	612190
尼日利亚 Nigeria	710331	734731
英国 United Kingdom	705245	702405
柬埔寨 Cambodia	608193	629950
伊朗 Iran	637779	695407
丹麦 Denmark	869066	670344
斯里兰卡 Sri Lanka	506636	519772
新西兰 New Zealand	432660	423123
土耳其 Turkey	431909	335326
法国 France	485342	505371
埃及 Egypt	344112	335613
荷兰 Netherlands	384476	370274
委内瑞拉 Venezuela	240780	284175
德国 Germany	261744	271185

10-11 主要沿海国家（地区）水产养殖产量
Aquaculture Production in the Major Coastal Countries (Areas)

单位：吨 (t)

国家或地区 Country or Area	2015	2016
世界总计 World Total	**76053701**	**80030862**
中国 China	47053234	49244101
印度 India	5260000	5700000
越南 Viet Nam	3438378	3624538
印度尼西亚 Indonesia	4342465	4950000
孟加拉国 Bangladesh	2060408	2203554
挪威 Norway	1380839	1326157
泰国 Thailand	928638	962571
智利 Chile	1045790	1035254
埃及 Egypt	1174831	1370660
缅甸 Myanmar	997306	1017614
菲律宾 Philippines	781798	796395
巴西 Brazil	574500	580500
日本 Japan	705452	676766
韩国 Korea, Rep.	479360	507962
美国 United States	426002	444369
中国台湾 Taiwan, China	313372	255183
伊朗 Iran	346118	398129
马来西亚 Malaysia	246205	201898
西班牙 Spain	289820	283828
尼日利亚 Nigeria	316727	306727
土耳其 Turkey	238964	250331
法国 France	163242	166140
英国 United Kingdom	211749	194492
加拿大 Canada	187374	200765
意大利 Italy	148763	157109
俄罗斯 Russia	152014	173104
墨西哥 Mexico	211562	221304

注：养殖品种包括鱼类、甲壳类、软体类等水生动物。
数据来源于《渔业和水产养殖统计年鉴2016》，联合国粮农组织。

Note: The cultivated varieties includes fish, crustacea, mollusc and other aquatic animals.
The data come from *Fishery and Aquaculture Statistical Yearbook,* FAO, 2016.

10-12 主要沿海国家（地区）石油主要指标（2017年）
Main Oil Indicators of Major Coastal Countries (Areas) (2017)

国家或地区 Country or Area	原油探明储量（亿桶） Crude Oil Proved Reserves (100 million Barrels)	石油存量（万桶）① Total Petroleum Stocks (10 000 Barrels)
中国 China	260	
孟加拉国 Bangladesh		
文莱 Brunei Darsm		
印度 India	46	
印度尼西亚 Indonesia	36	
伊朗 Iran	1580	
以色列 Israel		
日本 Japan		57600
韩国 Korea, Rep.		18400
马来西亚 Malaysia	36	
缅甸 Myanmar		
巴基斯坦 Pakistan		
菲律宾 Philippines		
泰国 Thailand		
越南 Viet Nam	44	
埃及 Egypt	44	
尼日利亚 Nigeria	370	
南非 South Africa		
加拿大 Canada	1700	19300
墨西哥 Mexico	73	5300
美国 United States	350	185600
阿根廷 Argentina		
巴西 Brazil	130	
委内瑞拉 Venezuela	3010	
法国 France		16800
德国 Germany		28900
意大利 Italy		11900
荷兰 Netherlands		12300
波兰 Poland		6000
俄罗斯 Russia	800	
西班牙 Spain		12100
土耳其 Turkey		6200
乌克兰 Ukraine		
英国 United Kingdom	26	7900
澳大利亚 Australia		3600
新西兰 New Zealand		830

注：①2014年数据。

数据来源于美国能源信息署。

Note: ①Data refer to 2014.

The data come from U. S.Energy Information Administration.

10-13 万美元国内生产总值能耗（2011年不变价，PPP）
Energy Use per Ten Thousand USD of GDP (Constant in 2011, PPP)

单位：吨标准油/万美元 (ton of oil equivalent per 10000 USD)

国家或地区 Country or Area	2011	2012	2013	2014	2015
世界 World	**1.34**	**1.32**	**1.30**	**1.27**	
中国 China	1.94	1.89	1.85	1.75	
中国香港 Hong Kong,China	0.42	0.40	0.38	0.37	
孟加拉国 Bangladesh	0.80	0.79	0.76	0.75	
文莱 Brunei Darsm	1.32	1.31	0.95	1.13	
柬埔寨 Cambodia	1.43	1.39	1.34	1.33	
印度 India	1.23	1.22	1.20	1.18	
印度尼西亚 Indonesia	0.95	0.92	0.89	0.88	
伊朗 Iran	1.57	1.74	1.74	1.79	
以色列 Israel	0.97	0.99	0.92	0.87	0.87
日本 Japan	1.05	1.01	0.96	0.93	0.91
韩国 Korea, Rep.	1.67	1.65	1.61	1.58	1.58
马来西亚 Malaysia	1.24	1.20	1.27	1.23	
巴基斯坦 Pakistan	1.13	1.10	1.09	1.06	
菲律宾 Philippines	0.74	0.74	0.72	0.72	
新加坡 Singapore	0.68	0.65	0.62	0.63	
斯里兰卡 Sri Lanka	0.56	0.55	0.48	0.48	
泰国 Thailand	1.29	1.29	1.35	1.33	
越南 Viet Nam	1.42	1.37	1.30		
埃及 Egypt	0.90	0.91	0.85	0.82	
尼日利亚 Nigeria	1.48	1.50	1.42	1.35	
南非 South Africa	2.25	2.17	2.11	2.18	
加拿大 Canada	1.80	1.73	1.83	1.83	1.76
墨西哥 Mexico	0.97	0.96	0.96	0.91	0.88
美国 United States	1.41	1.35	1.35	1.34	1.28
巴西 Brazil	0.91	0.93	0.94	0.97	
委内瑞拉 Venezuela	1.34	1.38	1.28		
法国 France	1.03	1.03	1.03	0.98	0.98
德国 Germany	0.90	0.90	0.92	0.87	0.87
意大利 Italy	0.78	0.78	0.75	0.71	0.72
荷兰 Netherlands	1.00	1.03	1.02	0.95	0.91
波兰 Poland	1.18	1.12	1.09	1.02	0.98
俄罗斯 Russian	2.24	2.22	1.99	1.92	
西班牙 Spain	0.83	0.85	0.82	0.79	0.80
土耳其 Turkey	0.86	0.87	0.71	0.70	0.71
乌克兰 Ukraine	3.34	3.23	3.06	2.98	
英国 United Kingdom	0.81	0.82	0.80	0.73	0.71
澳大利亚 Australia	1.35	1.31	1.27	1.22	1.25
新西兰 New Zealand	1.29	1.33	1.29	1.32	1.26

注：数据来源于世界银行WDI数据库。

Note: The data come from *World Bank WDI Database.*

10-14 主要沿海国家（地区）国际海运装货量和卸货量
International Maritime Freight Loaded and Unloaded in the Major Coastal Countries (Areas)

单位：万吨 (10000 t)

国家或地区 Country or Area	国际海运装货量 International Ocean Shipping Loading Capacity			国际海运卸货量 International Ocean Shipping Unloading Capacity		
	2000	2010	2017	2000	2010	2017
中国香港 Hong Kong,China	6770	11346	10697	10693	15428	17454
孟加拉国 Bangladesh	89	466	697	1408	3860	7951
文莱 Brunei Darussalam	10			102		
印度尼西亚 Indonesia	14153	50118		4504	11254	
伊朗 Iran	3065	4366		4486	8232	
以色列 Israel	1387	1927	2051	2920	2414	3743
日本 Japan	13010			80654		
韩国 Korea Rep.	15078			41882		
马来西亚 Malaysia	5483	11240		6922	13682	
巴基斯坦 Pakistan	617	1950		3080	4870	
新加坡 Singapore	32618	47146				
斯里兰卡 Sri Lanka	919					
埃及 Egypt						
南非 South Africa	2598			2598		
美国 United States	34334			83352		
阿根廷 Argentina	1550			1550	2374	
法国 France	6810	10421	11116	20273	20746	20764
德国 Germany	8602	10230	11566	14725	17070	17525
荷兰 Netherlands	9940	15335		32507	35617	
波兰 Poland	3152	3017	2831	1582	2838	4756
俄罗斯 Russian	828	20152		84	2353	
西班牙 Spain	5627			19343		
乌克兰 Ukraine	4271	8371		684	1744	
澳大利亚 Australia	48750	88736	146300	5418	8896	10106
新西兰 New Zealand	2214	3040	4061	1379	1798	2375

注：数据来源于联合国统计月报数据库。

Note: The data come from *UN Monthly Bulletin of Statistics Database.*

10-15 世界主要外贸货物海运量及构成（2016年）
World Major Maritime Freight Traffic in Foreign Trade (2016)

品种 Sort	海运量（百万吨） Freight Traffic (million tons)	
	2016	所占比例（%） Percentage
合计 Total	**11099**	**100.0**
原油 Crude Oil	1930	17.4
成品油 Refined Oil	1064	9.6
燃气 Fuel Gas	346	3.1
铁矿石 Ironstone	1411	12.7
煤炭 Coal	1108	10.0
谷物 Corn	473	4.3
其他货物 Others	4767	42.9

注：本表为估计数。

数据来源于《航运统计与市场评论》，2017年1月/2月，航运经济与物流研究所。

Note: This table is estimated data.

The data come from *Shipping Statistics and Market Review, January/February 2017, ISL.*

10-16 主要沿海国家（地区）国际旅游人数
Number of International Tourists of Major Coastal Countries (Regions)

单位：万人 (10000 persons)

国家或地区 Country or Area	入境（过夜）旅游人数 Number of Inbound Tourists (Overnight)			出境旅游人数 Number of Outbound Tourists		
	2000	2010	2016	2000	2010	2016
世界总计 World Total	**67732**	**95586**	**124496**	**82801**	**113724**	**145878**
中国 China	3123	5566	5927	1047	5739	13513
中国香港 Hong Kong, China	881	2009	2655	5890	8444	9176
中国澳门 Macao, China	520	1193	1570	14	75	125
孟加拉国 Bangladesh	20	30		113	191	
文莱 Brunei Darsm	98	21	22			
柬埔寨 Cambodia	47	251	501	4	51	143
印度 India	265	578	1457	442	1299	2187
印度尼西亚 Indonesia	506	700	1152	221	624	834
伊朗 Iran	134	294	494	229		901
以色列 Israel	242	280	290	353	427	678
日本 Japan	476	861	2404	1782	1664	1712
韩国 Korea, Rep.	532	880	1724	551	1249	2238
马来西亚 Malaysia	1022	2458	2676	3053		
缅甸 Myanmar	42	79	291			
巴基斯坦 Pakistan	56	91				
菲律宾 Philippines	199	352	597	167		
新加坡 Singapore	606	916	1291	444	734	947
斯里兰卡 Sri Lanka	40	65	205	52	112	145

注：数据来源于世界银行WDI数据库。

Note: The data come from *WDI Database of World Bank.*

10-16 续表 continued

国家或地区 Country or Area	入境（过夜）旅游人数 Number of Inbound Tourists (Overnight)			出境旅游人数 Number of Outbound Tourists		
	2000	2010	2016	2000	2010	2016
泰国 Thailand	958	1594	3253	191	545	820
越南 Viet Nam	214	505	1001			
埃及 Egypt	512	1405	526	296	462	
尼日利亚 Nigeria	81	156	189			
南非 South Africa	587	807	1004	383	517	
加拿大 Canada	1963	1622	1982	1918	2868	3128
墨西哥 Mexico	2064	2329	3508	1108	1433	2022
美国 United States	5124	6001	7561	6133	6106	
阿根廷 Argentina	291	533	556	495	531	1030
巴西 Brazil	531	516	658	323	646	853
委内瑞拉 Venezuela	47	53	60	95	148	153
法国 France	7719	7665	8257	1989	2504	2648
德国 Germany	1898	2688	3556	8051	8587	9097
意大利 Italy	4118	4363	5237	2199	2982	3085
荷兰 Netherlands	1000	1088	1583	1390	1837	1794
波兰 Poland	1740	1247	1747	5668	4276	4450
俄罗斯 Russian	2117	2228	2457	1837	3932	3166
西班牙 Spain	4640	5268	7532	410	1238	1541
土耳其 Turkey	959	3136	3029	528	656	789
乌克兰 Ukraine	643	2120	1333	1342	1718	2467
英国 United Kingdom	2321	2830	3581	5684	5556	7082
澳大利亚 Australia	493	579	826	350	710	993
新西兰 New Zealand	178	244	337	128	203	261

10-17 主要沿海国家（地区）国际旅游收入
International Tourism Receipts of Major Coastal Countries (Regions)

单位：亿美元 (100 million USD)

国家或地区 Country or Area	国际旅游收入 International Tourism Receipts						
	2010	2011	2012	2013	2014	2015	2016
世界总计 World Total	**10988**	**12495**	**12972**	**13811**	**14340**	**14370**	**13925**
中国 China	458	533	549	564	569	1141	444
中国香港 Hong Kong, China	272	337	380	426	460	426	380
中国澳门 Macao, China	227	390	445	523	516	320	306
孟加拉国 Bangladesh	1	1	1	1	2	2	2
柬埔寨 Cambodia	17	18	20	29	32	34	35
印度 India	145	177	183	190	208	215	231
印度尼西亚 Indonesia	76	90	95	103	116	121	126
伊朗 Iran	26	26					
以色列 Israel	55	60	62	65	64	61	64
日本 Japan	154	125	162	169	208	273	334
韩国 Korea, Rep.	144	175	197	193	230	191	211
马来西亚 Malaysia	182	196	203	210	226	176	181
缅甸 Myanmar	1	3		9	16	23	23
巴基斯坦 Pakistan	10	11	10	9	10	9	9
菲律宾 Philippines	34	40	49	56	61	64	63
新加坡 Singapore	142	181	193	191	192	167	184
斯里兰卡 Sri Lanka	10	14	18	25	33	40	46

注：数据来源于世界银行WDI数据库。

Note: The data come from *World Bank WDI Database.*

10-17 续表 continued

国家或地区 Country or Area	国际旅游收入 International Tourism Receipts						
	2010	2011	2012	2013	2014	2015	2016
泰国 Thailand	238	309	377	460	421	485	525
越南 Viet Nam	45	57	68	75	73	74	83
埃及 Egypt	136	93	108	73	80	69	33
尼日利亚 Nigeria	7	7	6		6	5	11
南非 South Africa	103	107	112	105	105	91	88
加拿大 Canada	184	200	207	177	175	162	183
墨西哥 Mexico	126	125	133	143	166	187	206
美国 United States	1680	1845	2001	2148	2208	2462	2447
阿根廷 Argentina	56	61	57	50	52	50	52
巴西 Brazil	55	68	69	70	74	63	66
委内瑞拉 Venezuela	9	8	9			7	6
法国 France	562	660	635	661	668	540	509
德国 Germany	491	534	516	552	559	474	521
意大利 Italy	384	454	430	462	456	394	404
荷兰 Netherlands	117	210	205	227	147	193	183
波兰 Poland	100	116	118	125	123	114	121
俄罗斯 Russian	132	170	179	202	195	133	128
西班牙 Spain	543	677	632	676	651	564	606
土耳其 Turkey	263	301	323	349	374	354	267
乌克兰 Ukraine	47	54	60	60	23	17	17
英国 United Kingdom	415	459	460	494	628	607	556
澳大利亚 Australia	311	342	341	334	341	313	345
新西兰 New Zealand	65	55	55	75	84	91	94

10-18 集装箱吞吐量居世界前20位的港口（2017年）
World Top 20 Seaports in Terms of the Number of Containers Handled (2017)

单位：万标准箱 (10000 TEU)

港　口 Seaport	所属国家或地区 Country or Region	吞吐量 Containers Handled
上海 Shanghai	中国 China	4023
新加坡 Singapore	新加坡 Singapore	3367
深圳 Shenzhen	中国 China	2521
宁波舟山 Ningbo Zhoushan	中国 China	2461
香港 Hong Kong	中国 China	2076
釜山 Pusan	韩国 Korea, Rep.	2047
广州 Guangzhou	中国 China	2017
青岛 Qingdao	中国 China	1831
迪拜 Dubayy	阿联酋 United Arab Em	1540
天津 Tianjin	中国 China	1507
鹿特丹 Rotterdam	荷兰 Netherlands	1370
巴生 Kelang	马来西亚 Malaysia	1198
安特卫普 Antwerp	比利时 Belgium	1045
厦门 Xiamen	中国 China	1038
高雄 Gaoxiong	中国台湾 Taiwan, China	1027
大连 Dalian	中国 China	971
洛杉矶 Los Angeles	美国 United States	934
汉堡 Hamburg	德国 Germany	880
丹戎帕拉帕斯 Tanjung Periuk	马来西亚 Malaysia	837
长滩 Long Beach	美国 United States	755

注：数据来源于上海航运交易所。

Note: The data come from Shanghai Shipping Exchange.

10-19 港口货物吞吐量居世界前20位的港口（2017年）
World Top 20 Seaports in Terms of the Cargo Handled (2017)

单位：百万吨 (million t)

港　口 Seaport	所属国家或地区 Country or Region	吞吐量 Cargo Handled
宁波舟山 Ningbo Zhoushan	中国 China	1009.3
上海 Shanghai	中国 China	750.5
新加坡 Singapore	新加坡 Singapore	626.2
苏州 Suzhou	中国 China	604.6
唐山 Tangshan	中国 China	573.2
广州 Guangzhou	中国 China	570.0
青岛 Qingdao	中国 China	510.3
黑德兰港 Headland Harbour	澳大利亚 Australia	505.3
天津 Tianjin	中国 China	500.6
鹿特丹 Rotterdam	荷兰 Netherlands	467.0
大连 Dalian	中国 China	455.2
釜山 Pusan	韩国 Korea, Rep.	401.2
营口 Yingkou	中国 China	362.7
日照 Rizhao	中国 China	361.4
光阳 Gwangyang	韩国 Korea, Rep.	291.8
烟台 Yantai	中国 China	288.2
湛江 Zhanjiang	中国 China	282.1
香港 Hong Kong	中国 China	281.6
南路易斯安娜 Southern Louisiana	美国 United States	279.4
南京 Nanjing	中国 China	236.4

注：数据来源于上海航运交易所。
　　数据为内外贸货物。

Note: The data come from Shanghai Shipping Exchange.
　　The data refer to the domestic and foreign trade cargo.

10-20 海上商船拥有量居世界前20位的国家或地区（2016年）
World Top 20 Countries or Regions in Terms of the Number of Maritime Merchant Ships Owned (2016)

国家或地区 Country or Area	艘数 （艘） Number of Vessels (unit)	总吨 Gross Ton （万吨） (10 000 tons)	载重吨 Deadweight ton	
			万吨 10 000 tons	占世界% Percentage in the World
世界总计 World Total	**52183**	**118269**	**177244**	**100.0**
巴拿马 Panama	6480	21652	33336	18.8
利比里亚 Liberia	3126	13633	21393	12.1
马绍尔群岛 Marshall Islands	2892	12795	20856	11.8
中国香港 Hong Kong, China	2420	10674	17242	9.7
新加坡 Singapore	2328	8070	12118	6.8
马耳他 Malta	2016	6565	9867	5.6
中国 China	3008	4790	7492	4.2
希腊 Greece	963	4226	7449	4.2
巴哈马 Bahamas	1160	5364	6806	3.8
英国 United Kingdom	757	2900	3890	2.2
日本 Japan	2620	2326	3378	1.9
塞浦路斯 Cyprus	797	2086	3300	1.9
挪威 Norway	788	1509	1907	1.1
印度尼西亚 Indonesia	2915	1219	1713	1.0
丹麦 Denmark	460	1500	1694	1.0
印度 India	850	960	1645	0.9
意大利 Italy	710	1550	1535	0.9
韩国 Korea, Rep.	1045	986	1465	0.8
葡萄牙 Portugal	352	1042	1371	0.8
德国 Germany	266	950	1032	0.6

注：按船旗统计，统计范围为300总吨及以上船舶。统计截止日期为2017年1月1日。
数据来源于《航运统计与市场评论》，2017 1月/2月，航运经济与物流研究所。

Note: According to the flag of the ship, the statistical range covers 300 tons ships and above.
The closing date of statistics was 1Jan, 2017.
The data come from Shipping Statistics and Market Review, January/February 2017 ISL.

10-21 集装箱船拥有量居世界前20位的国家或地区（2016年）
World Top 20 Countries or Areas in Terms of the Number of Container Ships Owned (2016)

国家或地区 Country or Area	艘数 （艘） Number of Vessels (unit)	载重吨 Deadweight ton	
		万标准箱 10 000 TEU	占世界% Percentage in the World
世界总计 World Total	**5147**	**1998**	**100.0**
利比里亚 Liberia	892	380	19.0
巴拿马 Panama	617	303	15.2
中国香港 Hong Kong, China	471	253	12.7
新加坡 Singapore	504	206	10.3
马耳他 Malta	285	134	6.7
马绍尔群岛 Marshall Islands	260	110	5.5
丹麦 Denmark	116	97	4.9
英国 United Kingdom	112	79	4.0
德国 Germany	117	75	3.8
葡萄牙 Portugal	165	63	3.1
中国 China	197	52	2.6
塞浦路斯 Cyprus	183	37	1.9
安提瓜和巴布亚 Antigua and Papua	238	32	1.6
美国 United States	61	22	1.1
法国 France	24	20	1.0
中国台湾 Taiwan, China	38	14	0.7
印度尼西亚 Indonesia	194	13	0.7
巴哈马 Bahamas	53	11	0.6
韩国 Korea, Rep.	89	10	0.5
卢森堡 Luxembourg	21	9	0.5

注：按船旗统计，统计范围为300总吨及以上船舶。统计截止日期为2017年1月1日。

数据来源于《航运统计与市场评论》，2017年5月/6月，航运经济与物流研究所。

Note: According to the flag of the ship, the statistical range of 300 tons and above ships. The closing date of statistics was 1Jan., 2017.

The data come from Shipping Statistics and Market Review, May/June 2017, ISL.

10-22 海上油轮拥有量居世界前20位的国家或地区（2016年）
Top 20 Countries or Regions in the World by the Possession of Ocean Going Oil Tankers (2016)

国家或地区 Country or Area	艘数（艘）Number of Vessels (unit)	总吨 Gross Ton（千吨）(1 000 tons)	载重吨 Deadweight ton	
			千吨 1 000 tons	占世界% Percentage in the World
世界总计 World Total	**14512**	**387847**	**636363**	**100.0**
马绍尔群岛 Marshall Islands	1229	58764	95573	15.0
利比里亚 Liberia	1012	47622	83337	13.1
巴拿马 Panama	1492	48174	80581	12.7
希腊 Greece	457	27807	49594	7.8
巴哈马 Bahamas	406	28243	43516	6.8
中国香港 Hong Kong, China	491	24202	42782	6.7
新加坡 Singapore	1004	25353	42382	6.7
马耳他 Malta	662	19523	32610	5.1
英国 United Kingdom	264	9043	14057	2.2
中国 China	748	8405	14057	2.2
挪威 Norway	221	7917	11880	1.9
日本 Japan	861	7711	11120	1.7
印度 India	150	5495	9626	1.5
印度尼西亚 Indonesia	674	4998	7723	1.2
意大利 Italy	241	4164	6878	1.1
百慕大 Bermuda	93	6872	6651	1.0
马来西亚 Malaysia	187	4438	5714	0.9
比利时 Belgium	46	3505	5428	0.9
科威特 Kuwait	29	2682	4918	0.8
丹麦 Denmark	163	3133	4742	0.7

注：按船旗统计，统计截止日期为2017年1月1日。

数据来源于按所有权模式计算的油轮船只总数，《航运统计与市场评论》2017年3月，航运经济与物流研究所。

Note: According to the flag of the ship, the closing date of statistics was 1 Jan., 2017.

The data come from the total tanker fleet by ownership patterns, SSMR March 2017, ISL.

10-23 海上散货船拥有量居世界前20位的国家或地区（2016年）
Top 20 Countries or Regions in the World by the Possession of Ocean Bulk Carriers (2016)

国家或地区 Country or Area	艘数 （艘） Number of Vessels (unit)	总吨 （千吨） (1 000 tons)	载重吨 Deadweight ton	
			千吨 1 000 tons	占世界% Percentage in the World
世界总计 World Total	**11139**	**424336**	**771086**	**100.0**
巴拿马 Panama	2474	105581	193200	25.1
马绍尔群岛 Marshall Islands	1199	51980	94838	12.3
中国香港 Hong Kong, China	1080	51154	93637	12.1
利比里亚 Liberia	943	43104	78964	10.2
中国 China	1087	28179	48257	6.3
新加坡 Singapore	547	26172	48086	6.2
马耳他 Malta	621	25186	45765	5.9
希腊 Greece	213	12680	23930	3.1
塞浦路斯 Cyprus	298	12009	21909	2.8
日本 Japan	439	10174	18963	2.5
巴哈马 Bahamas	268	9418	16601	2.2
英国 United Kingdom	104	7123	13458	1.7
韩国 Korea, Rep.	141	4701	8754	1.1
印度 India	103	2862	5176	0.7
意大利 Italy	59	2787	5138	0.7
葡萄牙 Portugal	51	2495	4538	0.6
挪威 Norway	70	2483	4350	0.6
土耳其 Turkey	80	2162	3745	0.5
印度尼西亚 Indonesia	230	2124	3597	0.5
菲律宾 Philippines	82	1809	3184	0.4

注：按船旗统计，统计截止日期为2017年1月1日。

数据来源于按所有权模式计算的油轮船只总数，《航运统计与市场评论》2017年4月，航运经济与物流研究所。

Note: According to the flag of the ship, the closing date of statistics was 1 Jan., 2017.

The data come from the total tanker fleet by ownership patterns, SSMR April 2017, ISL.